Western Lands and Waters Series
XX

Early hand-held hose and nozzle. Note bankwater in the background

HYDRAULIC MINING IN CALIFORNIA

A Tarnished Legacy

by
Powell Greenland

THE ARTHUR H. CLARK COMPANY
Spokane, Washington
2001

Copyright 2001 by
POWELL GREENLAND

All rights reserved including the rights
to translate or reproduce this work or parts
thereof in any form or by any media.

Arthur H. Clark Company
P.O. Box 14707
Spokane, WA 99214

LIBRARY OF CONGRESS CATALOG CARD NUMBER 00-040444
ISBN-0-87062-300-1

LIBRARY OF CONGRESS CATALOGING-IN-PUBLICATION DATA

Greenland, Powell, 1920-
    Hydraulic mining in California: a tarnished legacy / by Powell Greeland.
    p. cm. – (Western lands and waters series ; 20)
    Includes bibliographical references and index.
    ISBN 0-87062-300-1 (alk. paper)
    1. Hydraulic mining–California. 2. Gold mines and mining–California.
I. Title. II. Series.

TN421 .G72 2000
338.4'76223422'09794–dc21

00-040444

> I do not believe the Creator intends such
> ravage as the price of gold.
> *Thomas Starr King*

Contents

Preface	11
1. The Beginning	15
2. The Early Years	43
3. Water Ditches and Dams	71
4. Invention and Maturity	113
5. The Summit	159
The La Grange Hydraulic Mine	160
The Spring Valley Hydraulic Mine	182
[Malakoff Mine] The North Bloomfield Gravel Mining Co.	198
6. The Ending	227
Appendices	269
Notes	283
Glossary	295
Bibliography	303
Index	311

Illustrations

Early hand-held hose and nozzle	*frontispiece*
Map of hydraulic mining regions in California	14
Coyote Diggings as depicted by W. Pearson	25
An early hydraulic mine	35
Tertiary gravel channels, Nevada County	41
Small distributing box with four hose connections	51
Flume spanning a valley	87
Hanging a flume to a cliff	87
Bracket flume of the Miocene Mining Company	88
Craig's globe monitor	120
"To Hydraulic Miners" (advertisement)	123
"Hydraulic Chief" (advertisement)	124
Types of riffles	136
Hydraulic mining diagram in 1872 showing undercurrent	141
Ground plan and manner of firing a heavy blast, 1870s	155
The tailings sluice at the La Grange Mine	167
The "Castle" at the La Grange Mine	173
La Grange Mine	177
Spring Valley Hydraulic Gold Company Mine	185
Hydraulic diggings at Cherokee Flat as they appear today	197
The modern Bowman Dam	212
Undercurrents at the Malakoff Mine	215

The great pit of the Malakoff Mine 219
Hendy Giant on view at Malakoff Diggins State Park 225
The Marysville Levee Commission building 229
Greenhorn Creek showing ravages of the hydraulic mining . 245
Manzanite hydraulic mine 257
Debris-restraining dam below the Red Dog Hydraulic Mine . 264
Hydraulic mining debris dam sites 267

Preface

Anticipating the celebration of the centennial of the California Gold Rush by two full years, Rodman W. Paul's classic *California Gold* was the first in a stream of books dealing with some phase of that historic event to reach an eager reading public. His work was an initial attempt to put into historical perspective the California gold miner, in all phases of his occupation, and the mining of gold from the time of its discovery at Sutter's mill through its development as an industry until 1873. In the course of that study Dr. Paul devoted an important chapter to the subject of deep placers including both drift and hydraulic mining—surprisingly, the only significant work to appear on this form of mining during that prolific period.

Twelve years later, Robert L. Kelley wrote a definitive work on the debris controversy in the Sacramento Valley, *Gold vs. Grain*, which dealt with hydraulic mining in some detail, but only as a necessary adjunct to the central focus of his study. Finally, in 1970, two additional books appeared contributing more to the available pool of literature on the subject. In that year Jack R. Wagner published his popular *Gold Mines of California* with a chapter on hydraulic mining, including some excellent photographs, and historian Philip Ross May wrote *Origins of Hydraulic Mining in California*, a scholarly account of its early beginnings in the California mines. May, a New Zealander on study leave in California, noted in the course of his research that hydraulic mining ". . . became one of the best-known forms of California mining after the 1850s. Mining engineers, publicists, photographers and lithographers filled the pages of trade journals and popular periodicals with descriptions. Yet historians of the gold country have been sparse in their comments." He expressed bewilderment at the abundance of recent literature on the Gold Rush, Black Bart, Lola Montez, the Mother Lode, routes to the goldfields *et al,* yet not one monograph on the subject of hydraulic mining.

The present work, hopefully, will help fill that void.

This volume is in no way intended as a condemnation of hydraulic mining as it was practiced in California—nor is it an apologia. Although most of the research materials available for this study, such as government mining publications, were written by persons, for the most part, sympathetic to the continuation of the industry, it is hoped that sufficient objectivity and balance have been maintained to help the reader come to a fair judgment when arriving at the point in time of the Sawyer Decision in 1884.

The author's notes and selected bibliography, near the end of this volume, testify to the scope of the research materials utilized in this study, however, I would like to acknowledge invaluable help from the collection in the California Room of the State Library at Sacramento. The microfilmed copies of the *Mining and Scientific Press,* in particular, and the wide array of other publications were extremely helpful. The Milton manuscripts were an unexpected treasure. I would also like to express my thanks for the help I received at the local history branch of the County Library in Nevada City where endless hours were pleasantly spent reading from bound copies of the Nevada *Daily Gazette,* the *Daily Transcript* and the Grass Valley *Union.* The Marysville Public Library should also be acknowledged and thanked. Their fine collection contains much valuable information on the debris controversy, including a complete set of volumes containing the Sawyer Decision and copies of the *Daily Appeal* and other valuable Marysville publications. Also, the Research Center at Weaverville's excellent museum, operated by the Trinity County Historical Society, was most helpful in giving me access to manuscript materials that added much to my knowledge of the La Grange Hydraulic mine. I am extremely grateful for the help received from all of these institutions.

I would also like to express my sincere gratitude to Donald Duke and Vernice Dagosta for their patience and expertise in taking on the painstaking job of reading, correcting and smoothing my rough draft. Also, special thanks are due to Ed Tyson, Nevada County historian and librarian of the Historic Searls Library in Nevada City for reading my manuscript and for his cogent comments and encouragement. Finally, I am deeply grateful to historian Raymond W. Hillman of Eureka whose insightful observations have made this a better book.

POWELL M. GREENLAND
Port Hueneme, California
June 2000

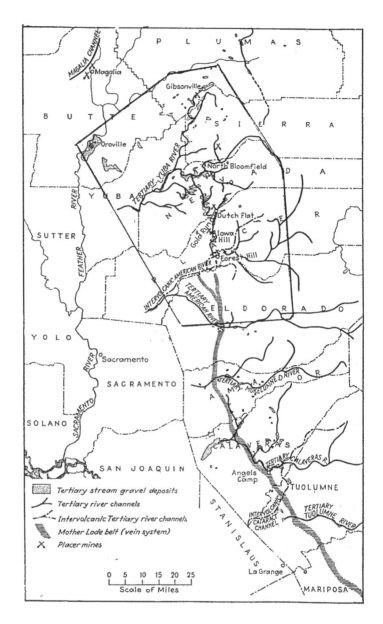

This composite map shows the main courses of the Tertiary river channels as mapped by Waldemar Lindgren together with the location of the Mother Lode and some of the leading placer mines from map by Olaf P. Jenkins. The enclosed area delineates the location of most of the principal hydraulic mines in California.

CHAPTER 1

The Beginning

Although a period of three months would elapse before James Marshall's gold discovery at Coloma, on January 24, 1848, was taken seriously, the floodgates then opened and an ever-increasing stream of Argonauts swarmed along the tributaries of the Sacramento and San Joaquin rivers in a frantic search for gold. By mid-year, mining camps stretched north to the rich diggings at Bidwell Bar and south to the auriferous sands of Mariposa Creek. So extensive was this movement that before the end of the year it is estimated that ten to twelve thousand men were scattered throughout the mines.

The river bars, and ravines, along those water courses soon became favored sites for scores of hopeful prospectors in their new-found quest. "What they call a Bar in California," one fledgling gold-seeker explained, "is the flat which is usually found on the convex of a bend in a river. Such places have always proved very rich, that being the side on which any deposit carried down by the river will naturally lodge, while the opposite bank is generally steep and precipitous and contains little or no gold."[1] The ravines, which led away from the streams, required the dirt to be hauled to the water for washing—hence the term dry diggings, came into general usage.

So with shovel and pick, on a promising bar, a small hole was excavated by the young miner, and the loose earth placed in a pan to determine the value of the location. Squatting beside the pool or stream, the pan was lowered beneath the water and the process of washing was carried to completion. The amount of gold left on the bottom was examined with care and if enough color showed to declare a panload—"pay dirt," this might just be the illusive spot where fortunes were made. "In the ravines is where most of the gold is found," exclaimed another lucky miner in '48. The loose stones and top earth being thrown off, the gravely clay that followed it was generally laid aside for washing, and the digging continued until the bottom rock of the ravine was reached, commonly at a depth of from one to six feet and often was found to be extremely rich. The dirt was then carried to the closest stream for washing.[2]

These earliest attempts were usually a process of trial and error as the neophyte miner learned his craft. But in a surprisingly short time, as Rodman Paul phrased it, these " farmers and clerks were quickly metamorphosed into miners."[3] Likewise these early methods of placer mining would inevitably evolve into the hydraulic mining process utilizing much of what had come before, beginning with the simple pan.

Gold pans could vary in diameter from ten to 18 inches, with sloping sides angled from 35 to 40 degrees. But even frying pans, with the handles removed, were often put to use in '48 and '49. Most miners preferred the larger size. The vessel, which was made of sheet iron was commonly filled about two-thirds full with dirt and placed under water. The contents were then stirred or kneaded with both hands to break up any large lumps and to free the gold from the clay-like material. At this time large stones were also removed. The pan was then grasped with both hands, slightly back of the center-line. In this manner it was held with a slight incline away from the operator. It was then raised until just covered with water, then given a slight oscillating, rotating motion, with the sand and gravel shaken from side to side. The lighter material was thus washed out of the pan with the heavier material, a mixture of gold and black sand, remaining in the bottom to be later dried and separated. This latter operation was often accomplished after supper when the material could be thoroughly dried over a fire and the black sand blown away by lung power. Sometimes, for this purpose, a "blower" was used which was nothing more than a shallow tin dish open on a portion of one side. The miner would then hold the blower with the opening away from himself, blowing out the sand and leaving the gold. If any black sand remained, it could be easily removed by means of a magnet.[4]

There appears to be a good deal of disagreement between experts regarding the number of pans a miner could handle in a ten-hour day. W. W. Staley, a mining engineer, wrote: "Most authorities agree that about one hundred pans of dirt are the most that can be panned by an experienced miner in a ten-hour day. On average, if the gravel weighs 135 pounds per cubic foot and the pan holds fifteen pounds this would be equivalent to about four-tenths of a cubic yard. With an 18-inch pan holding 25 pounds, this could amount to as much as one cubic yard in a ten-hour day."[5] To accomplish such a feat, only six minutes per pan could be allowed for each operation, including the time required to fill the pan with dirt and to remove the gold after the washing—quite a daunting ten-hour day. Historian John Walton Caughey, cuts this figure exactly in half. He contended: "To wash fifty pans was considered a good day's work, which would seem to mean ten or twelve minutes to the pan."[6] The best authority, of course, would be a working miner in the gold fields of '49, and one such, William Kelly, adds support to Caughey's assertion. "An expert hand, in anything like favorable ground, can gather and wash a panful every ten minutes; and the place that does not yield a quarter of a dollar to the panful is not considered worth working by that process, though it would give one dollar fifty cents per hour."[7]

Among the first wave of immigrants to reach the gold fields of California in 1848 and early '49 were a large contingent of Latin Americans. The first of these to arrive were Mexicans from the states of Sinaloa, Chihuahua, Durango and Sonora but by far the largest number was from the latter state and thus, the term "Sonorans," was applied to all Mexicans by the American miners during the gold rush period. Many of these men were professional miners, *gambusinos*, who had brought along their basic tool for placer mining—the batea. More cone-shaped than the traditional pan, the batea was made of wood and ranged from 15 to 24 inches in diameter with an angle of 150 to 155 degrees at the apex. The grain of the wood actually proved better for catching and holding the gold than the smooth metal pan. When available, bateas were eagerly sought after and highly prized by other miners in preference to the pan. Melville Atwood (also spelled Attwood), the well-known mining engineer and owner of the Grass Valley Assay Office, was so impressed with the batea he made a few improvements bringing it to a state of near perfection to aid in making tests where extreme accuracy was required.[8]

Perhaps the simplest form of placer mining, which ran its course in the early stages of the gold rush, was known as "crevicing" or "knifemining." This was practiced along the stream beds where cracks or crevices in the rocks, over the years, had become actually choked with gold. It then became a rather simple matter to remove the particles with nothing more than a knife.[9]

A major improvement over the gold pan, in terms of production, was introduced almost at the beginning of the gold rush. Isaac Humphrey, a placer miner from Georgia, arrived at Coloma as early as March 7, and within a week's time had introduced the "rocker."[10] Also known as a cradle, it was constructed of wood in the form of an oblong box with a bar across the middle. Rockers were designed in many forms and sizes and ranged in length from 24 to 60 inches and in width from 12 to 24 inches. The height varied from 6 to 24 inches. The bottom end of the receptacle was left open. A separate, removable box, designed as a hopper or sieve, occupied the upper or closed end of the rocker, the bottom of this device was made of sheet iron punched with holes three-eighths of an inch in diameter and located about one and one-half inches apart. Under the hopper and sloped toward the upper end was a canvas covered apron which retained much of the gold that fell upon it. The lower half of the box held the riffles which were made of horizontal strips of wood designed to catch the gold particles while the lighter sand and gravel was washed over the riffles and out the open end. On the underside, rockers similar to those on a child's cradle, were attached.[11] The simplest expedient for the construction of a rocker in '48 was simply a hollowed log.

Gravel was shoveled into the hopper while the cradle was rocked back and forth with a swift motion by means of a handle mounted on the side or closed end of the box. At the same time water was poured over the gravel or, better yet, a small stream was allowed to run over it. The miners soon learned it was better to keep an even flow rather than intermittent flooding. Two or three times, during a shift, operations were halted for the clean-up which consisted of scraping the gold and black sand from the riffles. When available, quicksilver was used to gather up and retain the gold particles.

Most miners followed a somewhat similar procedure as recorded in the journal of E. G Waite. "At noon the gold and black sand collected above the riffles were taken up on a scraper and thrown into a pan, which was carried to the river and carefully washed to remove, as far as possible, all but the gold. The yield of the forenoon was carried to the camp, dried over a blaze, the dry sand blown out and the gold weighed in scales, or guessed at, and poured into the partnership purse and deposited under the bed or anywhere else out of sight."[12] These appliances were so efficient that in the early months of the gold rush, even in their crudest form, rockers commanded a very high price. E. Gould Buffum, who found the cost well beyond his means, complained ". . . two hundred dollars being the price for which a mere hollowed log was offered us."[13]

The so-called "long tom" was an adaptation of the rocker and came into general use during the winter of 1849-1850 although it had been used in the gold region of North Carolina and Georgia for over 20 years. It consisted of a rectangular open-ended box about twelve feet long sloping in a downward direction. It was usually about eight inches in depth, with an upper end, one to two feet in width. From that point it widened so that at the middle it became about twice that width. The sides then continued parallel for the remaining length of the box. At the upper end, the bottom of the "tom," for a distance of six to eight feet, was made of perforated sheet iron much like the hopper in the rocker. It was designed to curve upward as it approached the end. Below the sheet iron the lower section formed the riffle box with wooden cleats running horizontally across the bottom. While in operation, water was run through the tom, either by a stream or, at a later period, a hose, to wash the dirt placed in the hopper. A man standing at the side of the box, stirred the dirt with a shovel, fork or hoe causing it to pass through the holes in the sheet iron. The larger stones were shoveled out of the box by the same operator. The finer material, which flowed through the perforations, was washed through the riffles and the gold was saved in the same manner as the rocker.[14]

The next improvement to be adopted by the California miner, was simply a trough made of boards one and one-half inches thick open at both ends, consisting of a bottom and two sides called a sluice box. It was generally 12 to 14 feet in length and 12 to 60 inches in width and 8 to 24 inches deep. Sluice boxes were designed in such a fashion that one end was two to four inches wider than the other (depending on the thickness of the board). The bottom boards were sawn for that special purpose. In this manner a continuous line of these boxes could be inserted into each other stretching out for any distance desired, from several hundred to several thousands of feet. They were constructed on trestles with a grade of from eight to 18 inches to every twelve feet. The degree of grade was determined by the nature of the material passing through the sluices. If the pay-dirt was heavy clay, for example, the grade would be made steeper to break up and dissolve the dirt; a 20-inch grade, however, was considered too steep. In very sophisticated operations the upper boxes would be constructed with a steeper grade to break up the mass, while the lower boxes had less grade for better recovery of the gold. As many as 12 to 40 men could be kept busy shoveling dirt into the sluices. A good days work could vary from two to five cubic yards of dirt per man. A continuous water supply covering the bottom of the sluice to a depth of two inches, was necessary for an efficient operation.[15]

Because the sluice box soon proved to be such an efficient method of placer mining, by the late '50s and early '60s, great care was given to its construction. When first adopted in California, sluices were often made without riffles and led into long toms for the purpose of saving the gold. However, this method was soon abandoned and the riffles in sluice boxes became progressively more sophisticated. The method of simply nailing horizontal strips of wood to the bottom of the sluice gave way to what John S. Hittell referred to as false bottoms. The most common was a longitudinal riffle-bar, which was from two to four inches thick, from three to seven inches wide and six feet long. Two sets of these riffle-bars would go into each sluice box, with the box twice as long as the bar, filling one-half of the box. They were wedged in, two inches apart. In this manner they could be easily removed for the clean-up operation.[16] The riffle-bars were usually sawn longitudinally with the grain. However, the so-called "block riffle-bars" were much more durable and were sawn in such a manner that the grain in the wood stood upright in the sluice box. Another type of riffle-bar was placed in the sluice box at a 45 degree angle from one side of the box to within an inch of the other side. Just below that bar another would be placed against the side about two inches below the open end of the first bar at a 45 degree angle in the opposite direction and continuing in this fashion, one after the other, to the bottom of the box. With this method much of the water and light dirt would run over the bars, but the heavier material, containing the gold, would follow the zigzag course. Quicksilver would be placed into the box by small steady drops, near the head of the sluice.[17]

At most sluicing operations a clean-up was conducted every six to ten days. The period of time between clean-ups was known as a "run." Since the clean-up was considered light work, it was often reserved for Sundays. If quicksilver was used in the placer mining operation, the gold was removed in the form of an amalgam which was a mass formed by the union of the gold, a solid, with the quicksilver, a liquid, becoming dough-like in its consistency sometimes a half an inch thick. The quicksilver and amalgam were then taken from the rocker, long tom or sluice and placed into a piece of buckskin or cloth and pressed so that the liquid would be squeezed out and the amalgam retained. During this early period, the amalgam was then placed in a pan or shovel and heated over an open fire to drive off the mercury. At a later period this operation would be accomplished in a retort and the mercury saved. What was left was a porous mass referred to as "sponge gold." Three pounds of amalgam could produce a pound of gold.[18]

It was a well-known fact in the gold fields that the long tom could double the production of the rocker per man and the sluice, double the production of the tom so, understandably, the sluice became the favored method of placer mining.[19] Writing from Double Springs, Calaveras County, Alfred Doten noted: "All sorts of improvements have been introduced in mining; the days of the 'tin pan and sheath knife,' are past and gone; the 'cradle' is thrown aside, and the 'long tom' is fast yielding to the sluice."[20]

From the descriptions of the various appliances which came into vogue, displacing the simple pan as the principal instrument for obtaining gold, it is obvious that the gold seekers found it necessary to band together in some sort of partnership arrangement to ply their trade. These groups could range from two to six or more miners and were frequently composed of men who had shared the rigors of the trail or shipboard life together and, in some cases, had come from the same or neighboring town. These groups were usually informal relationships organized from mutual need with each individual sharing equally the yield of a particular claim. A familiar theme running through the diaries or journals of these young men (mostly 20 to 35 years old) were the many instances where sickness or injury prevented a partner from doing his share of labor, but it was implicit that the ailing man would receive his equal share.

Although in 1849 and continuing through 1850 it was virtually impossible to find a miner willing to hire out for wages, as the claims became more scarce with the persistent influx of new arrivals, this labor shortage began to change. By 1854, as we shall learn, Mexicans were available for as little as $3.00 per day, Chinese for much less.[21]

Another form of placer mining, which had its beginning as early as July 1849, became known as river mining.[22] Because of the large labor force required and, as noted before, the scarcity of men willing to work for wages at that period, it was necessary to form stock associations with subscriptions paid for by labor. These more formal organizations were all the more essential because the period of labor, before any returns were realized, could sometimes be as long as five months. River mining involved the damming up and turning the flow of the volume of the river into a new channel by means of a dam. In this manner the gravel in the bed of the river was laid bare for working through a sluice. To accomplish this, a new channel or canal was excavated along the side of the riverbed diverting the stream and leaving the river bottom bare. If only a strip or portion of the riverbed was to be worked, a "wingdam" was constructed. It consisted of a water-tight wall which started from the bank out into the stream and then downstream in the form of an L. Any water that remained in pools or depressions was then raised either by hand pumps or water wheels and led back into the river.[23]

A more efficient method, which became popular in later years, was to actually divert the entire flow into a flume to expose the highly prized river gravel. A succession of water wheels, revolving on long shafts that extended over dugout pits and attached to pumps to keep them dry from seepage were often installed. As soon as the river was turned into the flume, the miners would begin digging up the bed, the gravel thrown into sluices and washed, while boulders were lifted out of the way with derricks and piled up on the bedrock that had been cleared away in advance.[24]

River mining was pursued at great risk. In a few hours an early storm could wipe out several months of back-breaking work with the season too far gone to start over again. Or, after five months of labor, the river bottom could be found barren of gold and the project abandoned. Although the stakes were high, the rewards could be worth the risk for fortunes were sometimes made from the gravel of a riverbed. Some early river mining was accomplished on a much smaller scale with gratifying results. "Within the last spring and summer some 15 points on this river [Middle fork of the American] have been dammed, the channel turned, and the bed of the river dug" wrote journalist E. Gould Buffum. "In one case, a party of five, dammed the river near what is now called 'Smith's Bar.' The time employed in damming off a space of some thirty feet was about two weeks, after which from one to two thousand dollars a day was taken out by the party, for a space of ten days. . . . Another party above them made another dam, and in one week took out five thousand dollars."[25]

Charles Ferguson, a young miner from Ohio was not nearly as lucky. After accepting an offer of a half share in a project to construct a wingdam at Long's Bar, he began his labor with great expectations. "When we had worked about four weeks on the dam, and had got nearly ready to rock the golden dirt," he later wrote, "a flood came as disastrous to our hopes as Noah's was to the ancient world, and swept everything away. I made up my mind that would be the last wing-dam, or any other kind of dam, that I would invest in California." He noted that at that time (1850) there was a "craze" for river damming. "There was one just below Long's, and another, the White Rock company's, where the dam and cutting, to turn the river, cost over a hundred thousand dollars that season, and they did not even get into the river. Besides at every turn or bend in the stream, there was a company wing-damming. In my opinion there was not one dollar got out of the river where ten dollars were put in."[26]

In some operations the lack of proper equipment such as water wheels and pumps could also lead to failure. As John W. Audubon observed, ". . . many men in parties of from twenty-five to fifty are here engaged in digging canals to drain the bed of the river at low water. I learn, however, that they are greatly hindered in this by numerous springs in the bottom of the river, and though there is no doubt a great deal of gold, the difficulties of getting it without machinery are more than can be realized."[27]

As the easily accessible stream-bed placers were becoming exhausted, prospectors broadened their search which included the higher elevations of the Sierra foothills. In the fall of 1849, a miner named Hunt discovered rich auriferous gravel deposits along Deer Creek, a tributary to the South Yuba, in what would later become Nevada County, and in a short time a mining camp was formed–aptly named Deer Creek Diggings (later Nevada City). While prospecting the bars and gulches along these water courses the miners noted that the canyons and ravines entering the streams immediately above the exceptionally rich bars often contained abundant quantities of gold dust. Upon following up and washing the gravel along these gulches,

THE BEGINNING

Coyote Diggings as depicted by W. Pearson.
Location on original not determined.

they discovered, on either side, what appeared to be rich streaks or "leads" which ran under the banks of the canyons.[28] When following such a lead in May of 1850, a rich strike was made on the slope of a hill northeast from town—destined to become the famous Coyote Lead. Word spread rapidly throughout the gold country, resulting in a stampede estimated, variously, from 6,000 to 16,000 miners during 1850 alone.[29] The location was soon named Coyote Hill derived from the curious method of mining performed at the site termed "Coyoting." An excellent description of the process has been preserved from the journal of Lorenzo Sawyer.

> The Cayota diggings are on the hills around the city; they are of great extent. . . . They are worked over by sinking a round shaft like an ordinary well, and the same in size. When the bed rock is reached, which is a rotten granite, the miners drift under in all directions. The gold is found on and near the bed rock. These are called Cayota diggings from a species of wolf which burrows in the earth. The Cayota diggings were commenced at the

foot of the hill and worked at first when the shafts were not more than six or eight feet deep. As they ascended the hill they were obliged to go deeper until they sank them to a depth of one hundred feet.[30]

The size of the claims were restricted to thirty feet with the lead proving to be one mile long and approximately one hundred yards wide. By 1856, Coyote Hill had produced in excess of $8 million. The town that sprang up near the site was appropriately named Coyoteville. Regardless of the richness of the gravel, only one in fifteen miners even made wages (eight dollars per day) from their efforts.

Peter Decker, a miner on Coyote Hill recorded this entry in his diary dated Wednesday September 25, 1850. "A town is laid out on Bucky Hill [Buckeye Hill] Kiota Diggins called Kiotaville. . . . Singular towns building country, hardly enough land to build all the towns on required here."[31]

Our diarist was also one of the many who failed in finding pay dirt on Coyote Hill. With a partner, John Rogers, the two sank several shafts but with no success. They finally resorted to hauling dirt down to the stream with a mule and wagon they had purchased where it could be washed "for those who had struck it rich."[32]

On a trip to Nevada City at a little later date, J. D. Borthwick also visited Coyote Hill and penned another account: "There is a variety of diggings here, but the richest are deep diggings in the hills above town, and are worked by means of shafts, or coyote holes, as they are called. In order to reach the gold-bearing dirt, these shafts have to be sunk to a depth of nearly one hundred feet; which requires the labour of at least two men for a month or six weeks; and when they have got down to the bottom, perhaps they may find nothing to repay them for their perseverance."[33]

These so-called coyote diggings, upon further investigation, were determined to be deep water-worn channels, once the beds of swift-flowing streams which had entombed rich gold-bearing gravels. The miners fittingly named these channels, "Dead Rivers."[34] Diggings of this sort were not confined to Nevada County but were found in many other parts of the gold region. Thus a camp named Coyote Bar was established in Placer County on the North Fork of the American River; a Coyote Diggings was established in Plumas County; another by the same name was also in Tuolumne County, where, it was said, claims paid from fifty to one hundred dollars per day; another Coyoteville was established in El Dorado County, and still another was located near Downieville in Sierra County.[35] A mining camp simply called Coyote was located in Calaveras County and visited by John W. Audubon who recorded the following account in his diary under the date of March 23, 1850:

> The first year these diggings were worked, many large amounts of gold were dug here with little labor; the second year required harder labor for poorer results; and it is its early reputation that keeps it up though some holes are still paying well; I was told four, out of the fifty being worked. The largest amount taken in the time I have been here, two days, was found by five Englishmen, two pounds and three ounces; others are well content with an ounce a day and do not give up their holes if much less than that is the result of ten hours or more of work.[36]

As we have learned from Peter Decker's account, the gravel dug from the Coyote Diggings had to be hauled to the nearest stream to be washed through a long tom or sluice. To solve this problem, as early as March 1850, near Nevada City, a small ditch one and one-half miles long was dug from Mosquito Creek to Phelps Hill, and another was excavated from Deer Creek to the same location, both designed to deliver water to the sluices and long toms located on the slopes away from a convenient source of water. By September a nine-mile ditch was started by Charles Marsh from Rock Creek to Coyote Hill and completed in December of the same year at a cost of $10,000. According to Charles Ferguson, who worked for wages on Coyote Hill, "Marsh was the pioneer of the supply for the diggings round about. He made a large reservoir to hold the water nights and Sundays, selling it out at the rate of an ounce of gold a day to the first user, to the next below, half an ounce, to those lower down a further reduction, when at last it found its way into Deer Creek."[37]

Other ditches soon followed, in March of the following year, Thomas & Company began the Deer Creek Mining Company's ditch from Deer Creek to Gold Flat a distance of 15 miles, completed in early 1852.[38] Miner's ditches were hardly confined to Nevada County and were similarly constructed throughout the gold country.

Charges for water, during this early period, with notable exceptions, generally ran at $4.00 per day to operate one long tom at the very head of the ditch and scaled down to $3.00, $2.00 and $1,00 per day for recycled water at the very end of the ditch.[39] This system led to a method for measuring and selling water known as the "miner's inch." While this standard varied somewhat throughout the gold region, the most widely used was determined by the amount of water that could pass through a two-inch plank with a one-inch orifice six inches below the surface of the water, measured over a 24-hour period. This quantity was calculated to be about 17,000 gallons or 2,274 cubic feet.

The water was measured through a "discharge box" which was designed with a long horizontal slit one-inch high and fitted with a slide to regulate the supply of water to be discharged. A "head of water" was the amount sold by the ditch company to any individual mining company. In the early days, this could vary from eight, ten or twelve up to 100 or more miner's inches. In later years, after the introduction of hydraulic mining, a "head" could be a thousand or even two thousand inches. The head was usually measured over a 24-hour period but it could be as short as ten or twelve hours and, in some cases, even six or eight hours. Water prices from 1850 to early 1853 ran as high as 75 cents an inch, but as the system of ditches grew the price fell rather substantially and even before the great water systems in the gold region were constructed, prices for water ran about 20 cents an inch or even less.[40] The costs of water, however, could vary rather dramatically depending on the supply in a particular area. Working at Spanish Gulch, Amador County, Alfred Doten, for example, made the following entry in his journal dated July 20, 1854: "Had a ten inch head of water—Afternoon washed out and had just an ounce—I have to pay 4 dollars for a six inch head and all over that is 50 cents an inch, so it costs me six dollars a day for a 10 inch head and as I have to pay 3 dollars for my man, I haven't made expenses so far."[41]

As the miners worked their way farther into the hills following prospects, they observed gravel deposits on the sides of the canyons and on benches of land, in some places hundreds of feet above the courses of the streams and rivers. During this period the miners learned that the richest gravels were those nearest to the bedrock while much of the over-burden, which seemed to be a volcanic material, had little or no value. At first these banks were worked by pick and shovel until, finally, someone got the idea of running a stream of water over the deposits to strip away the top layers of valueless dirt to expose the rich bottom layers. This method soon took the name "ground sluicing."[42]

Although C. A. Logan attributes the discovery of ground sluicing to an unknown miner in the California gold mines, the method actually dated back, at least, to the Middle Ages, if not to Roman times. Pliny the Elder, writing in the first century A.D. described in his *Natural History* a number of methods of placer mining later used in the California gold fields including an extremely dangerous process of undermining a hill and then washing the caved-in ground. Some historians have interpreted Pliny's description as ground sluicing others even claim it be a form of hydraulic mining (see appendix). But Georgius Aricola, celebrated as the Father of Mineralogy, in 1546 published *De re Metallica*, his magnum opus, a twelve-chapter treatise on mining and metallurgy, which included 292 carefully executed woodcut illustrations showing conclusively that all the methods of placer mining used in the California mines, including ground sluicing, had been in use before the 16th century. Mining historian Rodman Paul has suggested that, perhaps, it would be more correct to say that ground sluicing was reinvented in California.

James Mason Hutchings, a keen observer of life in California during those exciting times, wrote a concise description of ground sluicing just a few years after it had been introduced.

> The principle of operation is this, a bank of earth is selected which is desired to reduce or wash away, down to the pay dirt; a stream of water is conducted thereto, at so high a level as to command it; a small ditch is then cut along the portion to be ground sluiced, the water turned on and then any number of hands with picks and shovels either upon the edges of the ditch or by getting directly in the stream of water, pick away and work down the banks

and bottom, to be dissolved and carried away by the water, while the gold that may be contained in it, settles down without being conveyed or lost, to be finally saved by passing through the ordinary sluice.

Where the process is solely for the purpose of removing the top strata of earth in which no gold or pay dirt is found, down to that which will pay, it is called "stripping" by ground sluicing. Often, however when no pay is expected from the stripping process, the miner is unexpectedly cheered by finding in the top dirt more gold than sufficient to pay all the expenses of the operation.[43]

An interesting variation of ground sluicing, either invented or adopted in California, was known as "booming" or the "self shooter." It was actually a refinement of ground sluicing utilizing an automatic gate. Opening the gate caused a surge of water to be released, with great velocity, in a wave movement. This action was accomplished by means of a wooden gate attached to one end of a long sweep balanced on a fulcrum with a large wooden box or cistern attached to the other end. A small flume extended from the upper edge of a reservoir to this box. When the reservoir was full, the water flowed out through the flume into the box, which, upon being filled, moved suddenly downward raising the wooden gate allowing the water to rush out into a raceway leading to the mine.[44]

Another method of reaching the rich bedrock gravel beneath the layers of volcanic over-burden was a form of deep mining known as "tunnel" or "drift mining." This method was first introduced into Nevada County in 1850 and appears to have begun elsewhere at approximately the same time. Drift mining was adopted where the bedrock was more than usually enriched, and water scarce; or where, as was often the case, the gravel was covered by a heavy layer of volcanic rock either in the form of cemented layers or lava. To open a mine by drifting, a tunnel or adit was run into the hill, sometimes on the bedrock and sometimes below it and part of the gravel bed extracted. The roof was supported by pillars of gravel left untouched or by timbers. In the very earliest mines the gravel was removed by wheel barrows and dumped into a sluice, but in later years track was laid and the gravel removed by ore cars. With this type of operation, work could proceed the year around and if water was not available, due to the season, the extracted gravel was piled in dumps awaiting the winter rains.⁴⁵

Drift mining in the early '50s was a risky business. Miners with little or no knowledge of geology had a tendency to drive their tunnels above the bedrock completely missing the pay-dirt and ending up in barren layers of gravel. At a cost of five or six dollars per linear foot to as much as sixteen dollars per foot if tunneling through solid rock, this could be a very costly mistake.⁴⁶

Writing on this subject in recent years, Geologist and Mining Engineer Olaf Jenkins reflected: "The gamble of drift mining was caused by misconceptions of Tertiary physiography and millions of dollars were spent in vain. The geologic history and structure of the buried channels are so complex that the best of engineers have been baffled by them."⁴⁷

Tunnel or drift mining would soon become one of the principal placer methods in California and would remain so for most of the century. As early as 1861 in a single mining district, it was not unusual to find as many as fifty tunnels ranging from a few hundred feet in length to two or three thousand feet.⁴⁸ As techniques improved, drift mines became even more extensive in their underground works. For example, in 1878 it was reported, that the Bald Mountain Drift Mine near Forest City in Sierra County, which was the greatest producing tunnel mine in the state at that time, contained eight miles of drifts an underground railway system and a payroll of 250 employees. During a successive period of six years it produced over $100,000 annually.⁴⁹

It was ground sluicing, however, that was the catalyst that led to the only true invention of a completely new method of mining to be introduced in the California gold fields—hydraulic mining. In retrospect it would appear that hydraulic mining was a natural consequence of the improving technology of ground sluicing. Today it seems almost incredible that it took so long for someone to finally get the rather simple idea of using a hose and nozzle under pressure to wash down a bank of gravel. This conclusion appears especially true when it is borne in mind that ground sluicing had been in use, in Europe, 400 years before it was reinvented in California. As Rodman Paul commented: "The surprising feature of the Gold Rush is . . . that it took so long for the full measure of the Old and New World experience to be utilized."[50] To use a modern colloquialism the '49ers had to "reinvent the wheel" in the gold mines of California.

Antoine Chabot, a French Canadian, while working a ground sluicing claim in 1852, was the first miner to use a hose in a mining operation. The exact location of the site is still in dispute. It was either at Buckeye Hill located in the Greenhorn District, southeast of Quaker Hill in Nevada County; or as others believe, at Buckeye Hill Ravine, now actually within the boundaries of Nevada City, east of American Hill.[51] Wherever the location, Chabot brought water down a bank through a wooden box strengthened by metal straps. He was a sailmaker by trade so it is not surprising that he fashioned his hose from canvas, a material with which he was very familiar. The hose was reported to have been 35 feet in length and four inches in diameter. It should be noted that no nozzle was attached and it appears Chabot's purpose was simply to facilitate his ground sluicing operation.

In the spring of the same year, according to published reports, an unknown miner working auriferous gravel in a ravine at Yankee Jim's in Placer County, was the true inventor of the hydraulic mining operation. By installing a barrel on a 40-foot frame which was kept constantly filled by water passing through an elevated flume, a good head of pressure was maintained. At the bottom of the barrel a six-inch rawhide hose was attached, equipped with a four-foot nozzle made in the form of a tin tube tapered to a one-inch orifice.[52] This particular account received the authoritative sanction of Rossiter W. Raymond when it was published, from an article written by Charles Waldeyer, in the Fifth Annual Report of the United States Commissioner of Mining Statistics in 1873. The story continued: "This was the first hydraulic apparatus in California. Simple in design, dwarfish in size, yet destined to grow out of its insignificance into a giant powerful enough to remove mountains from their foundations." Raymond even included an illustration titled "Hydraulic Mining in 1852."[53]

In still another account, in a rather vague location, "near Marysville," Joe Wood from Ohio, a preacher's son, and a partner,

John Payne, a carpenter from Maine, sometime in 1852 invented hydraulic mining. A ditch carried water to a wooden box-like enclosure, one hundred feet long above the bank through a crude sailcloth hose connected to the box. The water was then brought to bear against the bank. The canvas hose also served as a nozzle, for, according to the account, it was tapered at the end for that purpose.[54] Historians give this version little acceptance because no authorities were cited in the story. Philip Ross May, who has done the most research into the origins of hydraulic mining in California, has found no supporting evidence confirming this story.[55]

Finally, the account that has received the widest agreement occurred at American Hill near Nevada City (now within the city limits), in the spring of 1853. The inventor was Edward Eddy Matteson (also spelled Mattison) who brought water to his claim, thirty or forty feet above a bank of gravel. The water entered a penstock or intake reservoir and from there through canvas hose four or five inches in diameter to a point in front of the bank of gravel. The hose, which was equipped with a sheet iron nozzle, was directed against the bank that was washed into the sluice boxes. Matteson used only 25 to 50 inches of water in this undertaking.[56]

The number of conflicting versions of the discovery of hydraulic mining caused Henry DeGroot to remark: "So obscure was its genesis that we cannot at this date determine just where or by whom it was first brought into general use, several different persons and places laying claim to this distinction—so many, in fact, and each with such apparent good showing, as to justify the suspicion that its introduction in these various localities might have been concurrent events."[57]

The first published account describing Matteson's invention was written by James Mason Hutchings in 1858 in a very popular little volume titled *The Miner's Own Book*. "By far the most efficient system of mining yet known, for hill diggings, is the Hydraulic; for the discovery of which California is indebted to Mr. Edward E. Matteson, formerly of Sterling, Windham County, Connecticut. Mr. M. first commenced the use of this method at American Hill, Nevada, in February 1852, and such was the success attending its operation the others around him immediately began to adopt it; and it is now in general use throughout the mining districts of the state."[58]

In subsequent years other accounts of Matteson's innovation have surfaced describing the hose, for example, as fashioned from rawhide, instead of canvas. Another interesting variation, credits the invention to a joint effort including, along with Matteson, Antoine Chabot and a former tinsmith named Eli Miller. The trio were supposedly partners, who had crossed the plains together on the trek to California. Through his skills as a metal worker, Miller fabricated a funnel to lead the water into the penstock and also the nozzle, described as made of tin, four feet long tapering to a one and one-half-inch orifice. Chabot, the sailmaker, of course was credited with making Matteson's hose. At still a later date, Matteson in his description of the event, recalled that his inspiration for using water against an embankment was prompted by the near miss of being buried alive by a falling bank he had loosened with his pick.[59]

Although Hutchings used the date of February 1852, as the month and year of Matteson's discovery and although J. Ross Browne also repeated 1852, as the year of this invention, (in his first annual report of 1867), most historians agree that it was actually a year later, in March of 1853. This date has been confirmed by the records of the Rock Creek Water Company which indicated that Matteson had requested the company to double his order for water from eight to sixteen inches on March 7, 1853.[60] Thus Edward Eddy Matteson has been designated the official inventor of hydraulic mining and a monument and plaque to that effect, commemorating the event, can be viewed today at the American Hill site, just inside the boundaries of Nevada City.

According to some accounts Matteson led the water from a supply ditch to a nail keg or pork barrel "set on a stump at the top of the bank."[61] This crude arrangement, according to an article in a Nevada City newspaper, led to the invention of the penstock, which it appears the writer felt, was as significant as any other part of his discovery. The reader will recall, however, that the penstock was first used by Antoine Chabot a year before Matteson's discovery. Nevertheless, the article gave a good description of this particular piece of equipment as it was used in hydraulic mining claims during the middle '50s.

> One of the old residents of Nevada City, Mr. E. E. Mattison conceived the idea of concentrating water in a tight compartment, and forcing it through a small pipe, thus making it to perform the duty of picking and shoveling its own gravel. The result was that the "penstock" was invented. It was a long wooden box made of two inch lumber, about 4 x 6 in the clear, well nailed together, and further strengthened by wooden clamps about 12 inches apart, or even closer, according to the pressure of the water. On the end of the "penstock" a hose, made of heavy duck, and terminating in an iron pipe, with about a half inch nozzle completed the apparatus, and it is thought to be the great invention of the age in mining, which it was . . . from this simple and rude primitive style of hydraulic mining."[62]

One of the best accounts of Hydraulic mining as it was practiced in 1858, just five years after its invention, was again written by California author, James Hutchings.

> Water was conveyed by canals and ditches around and among the hills and mountain sides where mining is carried on, it is thence distributed from the main canal through smaller ditches to the mining claims requiring it. Here it is run from the small ditch into a trough fixed upon trestle work,

Illustration from Rossiter W. Raymond's *Fifth Annual Report* showing an early hydraulic mine incorrectly dated 1852.

which is often technically termed the "Hydraulic Telegraph;" or run in heavy duck hose upon the ground, to the edge of the claim, thence over the edge and down the almost perpendicular bank to the bed rock or bottom of the claim, where it lies coiled about on the rock and dirt like a huge serpent. As the upper end of the hose is much larger than the lower end, the water running in, keeps it full to the very top; and the weight of the water, escaping through a pipe at the lower end of the hose, in a similar manner to that of a fire engine."[63]

Sometime in 1855 it was learned the auriferous gravel deposits that were being mined by either the tunneling method, or the new hydraulic process, were actually ancient riverbeds covered by volcanic debris. In time, miners and geologists were able to partially reconstruct their courses which had become veritable treasure caches with the gold-bearing gravel trapped below the over-burden near bedrock. These rivers dated back to the Tertiary geologic period at a time, during the Eocene epoch, when the newly discovered streams cut their courses. Over the passage of thousands of years, gold was liberated from quartz veins by prolonged erosion of the roof rock and the disintegration of the wall rocks under conditions of chemical decay. Gold, which is indestructible, became intermingled with gravels and washed into stream beds. During the later Miocene epoch, there followed a great outpouring of lava and volcanic ash covering these river systems and trapping their gravel deposits. At a later time, the Quaternary period, the Sierra Nevada was formed by the rising and tilting of the great mountain range. It sloped to the west with a sharp escarpment to the east. The new modern rivers, as we know them today, were then formed flowing in a westerly or southwesterly direction.[64]

Another theory that was current during the latter part of the 19th century speculated that California placer gold had been crushed and milled from quartz veins in the Sierra by means of glacial action. These particles were then washed into the ancient riverbeds.[65] It should be noted that modern geology does not completely accept this theory. As Olaf P. Jenkins stated: "It is a frequent fallacy of the placer miner to attribute the disposition of gold-bearing gravels to the action of glaciers. Contrary to such a belief, glaciers do not concentrate minerals ... [also] in Tertiary times they [glaciers] were wholly lacking."[66] In fairness, it should be noted that other authorities, such as Charles Scott Haley of the California State Mining Bureau, were not so positive. As late as 1923, Haley wrote: "The theory of pre-Tertiary glaciation has not been generally admitted, but it seems hard to explain the distribution of the gravels in this particular area [the San Juan Ridge] without admitting the probability of glacial action on these channels." He then went on to explain that over a period of from one million to five million years during the Cretaceous and Eocene periods there could have been climatic changes sufficient to support glaciers.[67]

In many cases, these ancient channels were cut by the modern or Quaternary rivers freeing the trapped gold which was deposited on bars and benches in their beds. It was this action of the newer rivers, cutting their way down into the beds of the older rivers that finally released the golden flakes found in Sutter's tailrace and all of the rivers and streams of the gold rush country. This was the gold mined by the '49ers and from these Quaternary deposits $500 million was extracted in a period of ten years.[68]

It now became apparent to the miners that there were three distinct sources of gold in California: Quaternary placer gold, distributed by the modern river systems, known to the miners as "Shallow Diggings;" and Tertiary placer gold, distributed by an ancient river system and buried beneath layers of debris, now called "Deep Diggings" and mined by either drifting or hydraulicking. And, finally, the gold associated with the Mother Lode—the gold-bearing quartz veins locked in the metamorphic rocks of the Sierra Nevada.[69]

The Mother Lode extends for 120 miles from Mariposa in the south to Georgetown in the north and varies in width from a few hundred feet to two miles. For the most part it parallels the Sierra system and varies in elevation from 1,000 to 3,000 feet. The term Mother Lode refers to an almost continuous series of gold bearing quartz deposits.[70] However, gold found in quartz veins is not confined to the so-called Mother Lode but extends in other systems comprising the great hard rock mining district at Grass Valley and others farther north. Altogether these gold fields, including the Mother Lode extend for a distance of over two hundred miles and are thirty miles wide. The mining of gold from quartz veins required the drilling and blasting of rock, deep beneath the earth, and then the crushing of the ore in *arrastres* or stamp mills in preparation for its separation. This method of extracting gold was aptly named "Hard Rock" or "Vein Mining" or "Quartz Mining."

In the early '50s it was learned, by both drift and hydraulic miners, that the upper stratum of Tertiary gravel deposits, for a depth of approximately one hundred feet, was loose and friable bearing a reddish color caused by the oxidation of the iron pyrites. The process of oxidation was helped by the action of surface water percolating through the gravel. Down lower, the stratum was found to be much denser and cemented showing a bluish color. This layer was also recognized as "pay-dirt" by the miners. In some areas a seam of pipe-clay was located between the two layers. These gravel deposits were also rich in vegetable matter such as sugar pine, manzanita, and other trees and shrubs. In the layer of pipe-clay, fossil leaves and ferns were also plentiful.[71]

The distribution of gold in a Tertiary channel could vary enormously. In general, in the upper levels, the sands and gravels were very poor. However, at the lower levels, within a few feet of bedrock, they began to get rich. On average the richest gravel was within three feet of bedrock. In the hydraulic mines the banks could vary in height from 50 to 500 feet or more. The top gravels could vary from two to ten cents per cubic yard. The average, from top to bottom, ranged from ten to 40 cents per cubic yard.[72]

Geologists determined that the occurrence of gold in paying quantities was limited almost entirely to the gravels in which quartz and metamorphic rocks formed the principal components.[73]

The richest Tertiary gravel deposits in California were located in the area drained by three major river systems—the American, Yuba, and Feather and also a smaller, but strategically located stream known as the Bear River. In Sierra, Yuba and Butte counties, the Tertiary channels were rich in gold-bearing gravels almost to the dividing ridge of the Sierra. In Nevada and Placer counties the extreme eastern parts were barren, however, the metamorphic area characterized by the so-called Washington belt and the Serpentine belt, which crossed both counties were extremely rich. In El Dorado County, the Tertiary rivers that cut across the Mother Lode were heavily charged with gold, but only on the down-stream side.[74]

In Amador, Calaveras and Tuolumne counties, in the very heart of the Mother Lode, the ancient river deposits were so heavily covered, according to Waldemar Lindgren, they did not lend themselves to hydraulic mining. However, there were small deposits in certain areas that were very rich, such as those near Chinese Camp.[75]

Tuolumne was the richest county in California in Quaternary placer gold output with a total production of $151,175,000. The Columbia basin, an open flat, with a radius of no more than a mile, produced more gold, for the size of the area, than anywhere else in the world—$55,000,000 from 1853 to 1870, yet this county was insignificant in Tertiary placer gold production. The Tertiary gravel deposits in Tuolumne County consisted of the Table Mountain Channel and the Tertiary Tuolumne River. But because they were so badly eroded by Quaternary or modern rivers and streams, they were largely barren or confined to pockets.[76]

Although the placers of Mariposa County produced a modest $10,000,000 all of this amount was taken from the stream beds and were virtually exhausted by 1860. All of the placer gold in this county was obtained from Quaternary deposits, which were both shallow and narrow with no remnants of Tertiary deposits remaining.[77] Calaveras was the only county in the Mother Lode with significant Tertiary gravel deposits. The Central Hill Channel ran from Altaville through Vallecito to Mokelumne Hill and another, the Fort Mountain Channel, ran southward from West Point nearly to Sheep Ranch then west toward San Andreas.[78]

In a study of the maps of the United States Geological Survey, Charles Haley indicated the principal Tertiary gravel deposits of the state were found in the counties of Sierra, Plumas, Nevada, Yuba, Placer and El Dorado. He noted that Amador, Calaveras, Tuolumne and Mariposa contained but small bodies of gravel, confirming the earlier studies. The greatest concentrations were found in the Slate Creek District (Plumas and Sierra counties), the San Juan Ridge District (Nevada County), and the Quaker Hill, Red Dog, Gold Hill and Dutch Flat region (Nevada and Placer counties).[79]

According to United States Commissioner of Mining Statistics, Rossiter W. Raymond, in his report of 1873, Placer County contained one of the most extensive auriferous gravel deposits in all of California. Beginning in the south at Todd's Valley and leading northward through Yankee Jim's, Wisconsin Hill, Iowa Hill, Indiana Hill, then turning in a northeasterly direction, to Gold Hill and Dutch Flat. It then entered Nevada County at Little York. The channel was from one-half to three-quarters of a mile in width with the deposits ranging from 75 to 500 feet in depth. Raymond reported that the area between the north and middle forks of the American Rivers was rich in auriferous gravel deposits, but he added: "Hydraulic mining was being retarded because of difficulty in bringing sufficient water to the area."[80]

The region in Nevada County known as the San Juan Ridge was considered by Waldemar Lindgren as the richest Tertiary gravel deposit in all of California. It extended from French Corral to Snow Point, a distance of 30 miles and was located on an immense ancient channel or riverbed that traversed the ridge of land embraced between the south and middle forks of the main Yuba River, coming from the direction of the Sierra Nevada Mountains and terminating at French Corral, the western extremity of the ridge. The ancient riverbed was from four to 600 feet higher in elevation than either of the two rivers defining its boundaries. The gravel deposits were from 200 to 1,500 feet in width and from 100 to 600 feet deep.[81]

From the time of the gold discovery until 1853—the year of Matteson's invention, the population of California had swollen from a sparse 14,000 to about 250,000. It is not surprising then that by the middle '50s the easily worked placers were becoming exhausted and throughout the gold fields the focus began shifting to the deep mines

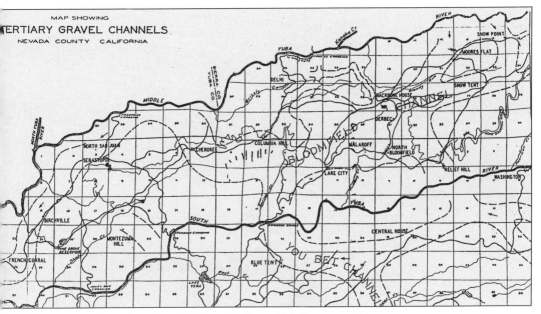

Map drawn by Edward C. Urgni, Bulletin 135, Division of Mines

of the auriferous Tertiary gravel deposits dominated by the new hydraulic process. The coming era would require the construction of a magnificent water system in the Sierra and thrust the mining industry into the forefront of engineering and technological advances—but the way would be plagued by trial and error, disappointment and considerable failure.

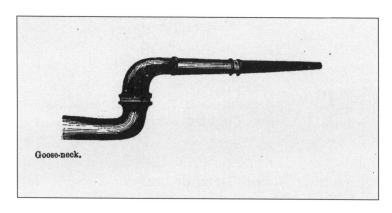

The Goose-neck nozzle.
[See page 44]
From August J. Bowie, Jr. *Hydraulic Mining in California*

CHAPTER 2

The Early Years

The earliest nozzle which was presumably attached to the duck or cowhide hose held in the hands of either E. E. Matteson at American Hill or an unknown miner at Yankee Jim's has been variously described as: made of brass or copper tapered to a diameter of from one to two inches; or an iron pipe with about a one and one-half-inch nozzle; or a sheet iron nozzle; or a four-foot tin tube narrowing to a one-inch-diameter; or constructed of tin plate tapered gradually in size from the flexible hose down to one or two inches in diameter; or fashioned with sheet brass enclosed in a wooden jacket with a three-quarter-inch-nozzle.[1]

Regardless of which of the above was actually used in the initial event is of little consequence. But the descriptions of these various nozzles are significant to the historian for each describes a type actually in use in 1853 and, perhaps, for some years later.

The first significant improvement to these primitive nozzles was a cast brass pipe four or five inches at the butt and connected to the hose by means of a screw joint. This pipe was too heavy to be held by hand and was, therefore, strapped on a tripod or secured to the top of a post, firmly anchored in the ground, or even fastened to a convenient tree stump.[2]

As water pressures were increased by raising the elevation of the cisterns located above the mine sites, improvements were needed to bolster the bursting pressure of the duck hose then in general use. The strongest duck or four-inch copper-riveted leather hose would not hold more than 80 feet of water and would be worn out after just a few months use. Neophyte hydraulic miners, working in remote locations, soon became proficient in the repair of ripped canvas hose and others, in its fabrication. As early as 1854, in mining camps that catered to their needs, bolts of number 1 duck became a standard commodity.

To strengthen these early hoses, as many as three layers of canvas were used in its manufacture, then rope binding or strong netting was wrapped around the outside giving the hose added strength, while also increasing its life-span. At a little later period the hose was surrounded with galvanized iron bands, two inches in width and from one to three inches apart. They were connected together with four ropes running longitudinally from one band to the other giving the appearance of crinoline which was a reference to ladies' stiff petticoats and hoop skirts then in fashion. Crinoline hose, as it was promptly dubbed, could resist pressures of 180 feet or 75 psi.[3]

In the latter part of 1853, the first iron pipe was introduced into the hydraulic mines of California by R. R. Craig at American Hill, near Nevada City. This innovation, although amounting to nothing more than 100 feet of stove pipe, was so successful it came into general use.[4] Pipe, however, could not be attached directly to the nozzle because a flexible joint was needed to allow for movement. To solve this problem a short length of hose was installed between the sheet iron pipe and the nozzle.

Within the short time frame of two years another improvement was available with the invention of the "gooseneck" nozzle by J. M. Allenwood of Timbuctoo, Yuba County. Engineered to eliminate the need for the flexible link it was designed with two elbows, one working above the other, with a coupling joint between them. However, this device proved to have serious problems. Under pressure the joint moved with great difficulty. When turned in a horizontal direction it sometimes "bucked" or spun around in the opposite direction. The same problem occurred when it was operated vertically.[5] The maximum operating distance was a spectacular 250 feet, but beyond this point the column of water tended to break up and lose its efficiency. It was soon learned that for smooth operation it had to be firmly secured to the ground and turned in either direction by means of a block and tackle.[6]

It appears these were not the only problems experienced by miners using the gooseneck nozzle. From the pages of the Nevada City *Daily Gazette* we learn of an accident that occurred in late June 1868: "A miner working a claim at Blue Tent while adjusting his hydraulic nozzle under full pressure, the gooseneck became detached from the pipe. The miner was struck by the full force of the water which knocked him into the sluice-way and he was washed some distance. Although severely bruised his injuries are not considered life threatening."[7]

But despite these problems, the gooseneck came into moderate use because it could operate with much greater volume causing banks to be washed away in an amazingly short time leading the way for greater improvements in years to come. It was after the introduction of this latest innovation that the term "hydraulicking" was added to the miner's lexicon.[8]

Although the gooseneck nozzle found growing acceptance, particularly in larger operations, and iron pipe was in general use as a conduit, it should be noted that as late as 1863, John S. Hittell in describing the "piping" at a typical hydraulic mine wrote: "The pipe at the end of the hose is like the pipe of a fire engine hose, although usually larger. Sometimes the pipe will be eight inches in diameter where it connects with the hose, and not more than two inches at the mouth; and the force with which the stream rushes from it is so great, that it will kill a man instantaneously and tear down a hill more rapidly than could a hundred men with shovels. One or two men are required to hold the pipe. They usually turn the stream upon the bank near its bottom until a large mass of dirt tumbles down, and then they wash this all away into the sluices."[9]

For an understanding of the evolution of hydraulic mining in California it must be stressed that the new process did not emerge full blown after its invention in 1853. During its brief 30-year life-span in the foothills of the Sierra Nevada before production was severely curtailed, 20 years would pass before the new method would reach maturity and achieve the stage of development it enjoyed at the time of the Sawyer Decision in 1884.

Commenting on the development of hydraulic mining after Matteson's discovery at American Hill, one historian exclaimed: "The results were immediate and profound, an entirely new industry was born. Wherever Tertiary gravels were being worked, rudimentary water systems were hastily enlarged, claims were being fitted up with hoses and nozzles, and the roar of hustling water filled the air."[10]

The enthusiasm of this appraisal should be balanced against the less dramatic assessment of historian Rodman Paul: "Despite its acknowledged efficacy, the hydraulic process did not pass into general employment until several years had elapsed."[11] Henry DeGroot left this more contemporary perspective: "As the discovery of this method was due to accident so has its subsequent growth been the creature of evolution. From the smallest beginnings it grew little by little every year, bringing with it some improvement and adding something to its efficiency, until it reached vast capabilities and gigantic proportions."[12]

The application of profitable hydraulic mining was restricted by a number of limiting factors. Assuming, of course, the claim was located on an auriferous gravel deposit, first and foremost was the necessity of a generous water supply. Over a year after the invention of hydraulic mining much of the finest auriferous gravel deposits in the gold region were left untouched because of the lack of water. The Sacramento *Daily Union* sounded a growing concern: "There are thousands of acres in the state of surface diggings . . . which will pay good wages whenever water can be obtained to wash the dirt. It is found that gold is scattered in rich profusion over the thousand hills which lie scattered between the plains and the base of the Sierra Nevada."[13] Also the grade at the claim needed to be adequate for the easy disposal of tailings through the sluices. In the ensuing years those areas which contained the most extensive districts suitable for the application of hydraulic power became the most prosperous, while the older towns dependent on river mining or on shallow placers fell into decay and were partially and in some cases entirely deserted.[14]

To determine the desirability of an investment in a hydraulic mine, after the value of the gravel was ascertained and water located, it was next necessary to legally claim the rights to the water then survey the route of the water course from its source and the specifications and costs for dams and reservoirs, ditches and flumes. In making these determinations sufficient fall or grade was required to develop the head or pressure needed to work the gravel. All of these conditions were found to be present in the vicinity of Deer Creek and the South Fork of the Yuba River, actually within the city limits and the outskirts of Nevada City, the site of Matteson's initial discovery. It is not surprising, therefore, that one of the very earliest hydraulic mines in California grew out of the consolidation of four early coyote claims known as the Manzanita Diggings. At this location, shafts had been sunk to bedrock and the gravel drifted out by hand using a windless and whim. After consolidation a bedrock tunnel was driven for drainage in preparation for hydraulic mining and the claim became known as the Tomlinson Diggings. In later years, however, the name would revert back to Manzanita.[15]

Another district which proved very suitable for hydraulic mining, lay in the area of Yankee Jim's in Placer County. Here the new method was introduced within two months after its invention by Colonel William McClure who had heard of Matteson's new method and had actually visited the location at American Hill. Nearby, a close neighbor, Iowa Hill, was the site of another very early hydraulic mine. In fact one of the earliest known illustrations of a hydraulic mine is pictured on a letter sheet titled "Rich claim, Iowa Hill, 1853." The representation shows a mining operation on a heavily worked hillside with pipe leading from a flume at the center of the hill-top.[16]

Commercial fire hose, although quite expensive, was soon a popular item in the California gold fields. The Placer *Herald* informed its readers that the Jamison Company, also at Iowa Hill, had purchased 350 feet from a San Francisco supplier for the sum of $13,000 to carry water from a reservoir two hundred feet above the claim through a two and one-half-inch nozzle. The article noted that the pipeman took a position 30 feet from the bank playing the stream against the gravel. The four-inch hose was wrapped with strong cord for added strength. The writer added, "the stream of water will do more digging than twenty men."[17]

At still another claim, in the same locality, a cut was excavated into a hill 120 feet deep where two lines of hose and two nozzles were used. The pay dirt was only about ten feet above bedrock making it necessary to wash off 110 feet of ground. To accomplish this feat, two sets of sluices were installed—one for the pay dirt, and the other to carry off the worthless gravel. Two men operated the pipes standing 20 to 30 feet from the bank. It was reported that "the force is so great that the streams of water force their way, as if by magic, into the bank diggings."[18]

Matteson's invention met with immediate success at Indian Hill in Sierra County where a piece of ground 40 feet wide by 80 feet long, using 200 inches of water, was "piped off" in just six days. The company employed ten men at $4.00 per day which totaled $240 in wages. Three hundred dollars was spent for water at the rate of 25 cents per inch, one hundred dollars more for quicksilver and various maintenance costs. During the ten days they grossed $3,000 leaving a net profit of $2,350. The total amount of gravel washed during this time was 224,000 cubic feet.[19]

At Eureka City, also in Sierra County, just north of Goodyear's Bar, it was reported by the *Sierra Citizen* that hydraulic mining had readily been adopted in that remote location. "The miners here are doing well; the claims are all open and you can hear nothing but the roaring of water, the rattling of gravel and the never to be forgotten grating of the sluice fork . . . The Smith boys have the best claims in this locality—last week three men took out $700. One claim adjoining, sold, two days ago, for $1,000 cash." [20]

Farther north in Butte County a number of mines in the vicinity of Bidwell Bar, by 1854, had begun hydraulicking, employing both canvas duck and rubber hose with one-inch nozzles. While in neighboring Shasta County the first hydraulic mining was introduced in the Horsetown area with such success that, by 1856, eleven mines were in operation yielding from $500 to $800 per week. Other early hydraulic camps in the county were located at Texas Springs and Squaw Creek.

The new method first appeared in Plumas County, near the mining town of Rabbit Creek (later La Port), as early as 1853, where a crude one-inch nozzle was in operation with such success that by the end of the year 50 companies were working along Rabbit Creek. Two years later, with the construction of the Martindale Ditch, hydraulic mining became even more prosperous. Meanwhile a number of primitive, two-man hydraulic operations were attempting to just make wages, with minimal success, in the Nelson Creek-Rich Bar region of Plumas County.[21]

Hydraulicking rapidly spread into the southern mines shortly after its invention. On September 10, 1853, the Calaveras *Weekly Chronicle* under the heading, "HYDRAULIC METHOD IS INTRODUCED INTO MINES" reported that the revolutionary new method had been adopted by Major Case on his claim at Stockton Hill in Chili Gulch. "A powerful stream of water is brought to bear on the bank. Major Case estimates the hose does the work of ten men."[22]

In nearby Tuolumne County the Columbia *Gazette* enthused, ". . . hydraulic washing has commenced in this gold mining district with great success. The abundant supply of water from the Tuolumne Water Company have enabled miners to turn to this new labor-saving method. Big piles of dirt and hillsides are disappearing, and heavy clay deposits are washed into the sluices. On the Broadway road a head of water has been elevated about thirty feet and played on dirt ordinary washing wouldn't effect."[23] It should be noted that it was in Tuolumne County near Columbia in October 1855 that miners first made the important geological discovery that the Tertiary gravel deposits were actually ancient river beds.[24]

With the completion of the Johnson Ditch in Amador County,

hydraulic mining was first attempted with moderate success at the mining camp of Muletown near Ione. One of the earliest claims, the Monumental, paid $8 to $15 per day to each of the four men working the mine. At a neighboring claim, 36 ounces of gold were obtained at a single clean-up. The price of water charged by the Johnson Ditch Co. at Muletown, at this early period was 33 cents a miner's inch.[25]

In the winter of 1853-54 hydraulic mining was also firmly established in the vicinity of Volcano. The introduction of the new method at this location was a simple and logical extension of prior mining practices. Two years earlier, after the construction of the Johnson and Volcano Canal companies' ditches, extensive ground sluicing had been profitably conducted in the area. Both ditch companies took their waters from different forks of Sutter Creek.[26]

During the first ten or twelve years after the introduction of hydraulic mining, water was often brought to a mine site from a company ditch into a 16-inch to 20-inch pipe above the bank near the center of the claim. From that location it was discharged into a smaller pipe, laid at right angles, which, in turn, carried the water in both directions to connections at either end of the pipe. From the top of the bank the water was then conducted through two distributing pipes directly to the nozzles.[27] A more common method was simply to construct a small reservoir or cistern at the top of the claim, sometimes in excess of 100 gallons, and lead a crinoline hose over the bank capped with a nozzle.[28]

After 1860, at larger more sophisticated operations, water was conducted from the ditch or canal through a flume to the head of the mining ground at an elevation of from 100 to 200 feet above the bedrock, from there it was conveyed to the bottom in iron pipes often supported by an incline of timber. These pipes (usually two in number) led into a cast iron box, located at the bottom of the claim. This receptacle, prismatic in shape, was known as a "distributing box" which was equipped on the top and sides with four or as many as six connections with flexible crinoline hoses attached terminat-

Small distributing box with four hose connections. Located at Malakoff Diggins State Park, North Bloomfield, California. *Author's collection.*

ing in metallic nozzles. This appliance was the invention of Marcos A. Winham of North San Juan, who, in turn, licensed its manufacture to the Smith & Low Foundry in the same town. The complete unit actually consisted of three separate patented items: the iron box; a coupling, for the purpose of easily attaching the hose; and a gate valve, installed at the end of the distributing pipe where it connected to the box. Before the invention of the gate valve, according to an article in the *Mining and Scientific Press*, the operator "must necessarily ascend a perpendicular ladder or slippery steps cut zigzag in the steep bank, often to the height of 150 feet, clad in his cumbersome gum boots and waterproof suit, in order to turn the entire current of water off."[29]

With the spread of hydraulic mining in the mid-'50s, the rich gravel deposits on the San Juan Ridge, situated between the Middle and South Forks of the Yuba River, became favored locations for dozens of new hydraulic claims. One of the earliest water companies serving this region was the Grizzly Ditch Co. organized in 1853 to bring water exclusively to the mines of North San Juan. But the growing demand for water would not be satisfied until the completion of the Middle Yuba Canal Company's ditch, also begun in 1853, taking water from the river at a point just below Woolsey's Flat. At this location the company dammed the stream and constructed a reservoir designed for a capacity of 1,500 inches. By 1865, after this small beginning, the Middle Yuba Canal Company had expanded its operations to reflect a total investment of about $600,000.[30]

A few of these early ditches, with notable exceptions, proved to be quite profitable. Water sales for the Middle Yuba Canal Company, from January 20, 1856 to July 1, 1864, were a respectable $968,033. But net earnings were even more impressive. From January 1, 1860, to January 1, 1863, the total was $237,043 and from July 1, 1863, to year-end, $128,959.[31]

Geologist and mining expert Benjamin Silliman, who was very familiar with the buried wealth in that region and the ever-increasing need for water to release its bounty, reflected: "So far as my knowledge of California affairs extends, there is no single franchise comparable in intrinsic value and immediate returns with the ownership of the united rights of the Middle and South Yuba."[32]

By 1857 there were 696 miles of ditches in Nevada County alone reflecting a total cost of $1,500,000. Throughout the entire gold region of California by 1867, there were a total of 5,328 miles of main channels with at least 800 miles more of branch ditches representing an investment of $15,575,400.[33]

Shortly after Matteson's invention, the new method spread rapidly through a good portion of the gold regions of California where ancient gravel deposits were abundant and particularly the areas drained by the branches of the Feather, Yuba and the American rivers which encompassed El Dorado, Placer, Nevada, Yuba, Sierra and Butte counties. A new era of prosperity seemed firmly established. But surprisingly the apparent bonanza from this great California-born invention proved short-lived. After about five years of feverish activity, the new industry actually took a down turn. In all parts of the gold country where hydraulic mining had been adopted, because of limited technology and the inexperience of the hydraulic miners, the working of claims had accelerated in a few years to the point of saturation.

By 1857, in some of the richest hydraulic mines, it was learned that below varying depths, depending on the location, the easily worked gravel became so solidly packed, that in many locations, it had the characteristics of "cement." This material was impervious to the small nozzles and limited water pressure then available. Compounding matters even more, it was found that many of these new claims were too restricted in size to support the substantial investment needed to properly develop the property. Small claims would be worked out in a matter of weeks, or at best a few months. It was learned that from 80 to 200 acres were needed to justify the capital investment required.

Also, an enduring problem that plagued the hydraulic miner, and fundamental for success, was the water supply. In too many locations the essential quantity was simply unobtainable. Other problems haunted the hydraulic miner as well. The working pit often reached a point where the gravel was devoid of an outlet and large expenditures were necessary to construct costly bedrock cuts or drainage tunnels. As Charles Waldeyer explained: "In the principal hydraulic mines in the State early operations were commenced from the easiest point of access, without much regard to the lasting sufficiency of the 'open cut' or tunnel. The natural consequence was that only a small portion of the mine could be worked by existing facilities; the grade had to be cut down from time to time, and in most cases the constant improvements absorbed or exceeded the whole production of the mine." The day of the small operator with limited capital was over. Thus, after five or six years of quick prosperity, the industry actually suffered a relapse and scores of claims were abandoned or sold off at a fraction of their worth.[34]

Commenting on these conditions Samuel Bowles observed:

"The maps of five years ago shows [the slopes of the Sierra foothills] crowded with names of mining villages, while all the rest of the state seemed bare of settlement; but now half of these are wholly deserted, and the others, with few exceptions, are in a decaying condition, with many houses and stores unoccupied, and often with only a small proportion of their old populations."[35]

Rossiter W. Raymond, in his *Report* of 1869, included a letter from correspondent W. A. Skidmore who noted that the towns of Hunt's Hill, Red Dog, You Bet and Little York, all in Nevada County, formerly had been the centers of active hydraulic operations, "but now this kind of mining [hydraulic] was laboring under a depression incidental to the business throughout the state." He reported the town of Red Dog had half its buildings unoccupied and none of the cement mills were in operation. Little York had deteriorated nearly as much. Dutch Flat, in Placer County, had lost half its population. Once listed at 2,000 it had been reduced, according to Skidmore, to barely 1,000. On the main streets he found dilapidated sidewalks, vacant hotels and stores and ruined houses. "This part of the country," he commented. "is now passing through a crisis . . . mere labor has nearly exhausted its efforts, and capital has not yet come to its assistance. The gold is in the ground, but not so near the surfaces to be extracted by mere labor."[36]

The gold camp of Alpha in Washington Township was at the height of its prosperity from 1854-1855 supporting a large number of productive hydraulic mines and said to be one of the liveliest towns in Nevada County. However, the easily worked diggings were soon exhausted and by 1867 only one mine was still in operation. Alpha was then a ghost town with only memories of the days when over $1,250,000 in gold had been recovered from its gravel deposits.[37] Relief Hill on the rich San Juan Ridge was also a casualty of this technological syndrome. First settled in 1853, with the advent of hydraulic mining, intensive operations were very widespread and by 1858 the town could boast of over 100 voters. In a few years, however, the camp was nearly deserted.[38]

An article in the *Mining and Scientific Press* asked the question,

"What is the future of our mining towns? Who does not ask the question yet half dread to hear it answered?" The article then continued: "At this moment there are hundreds of business men speculating on it; and hundreds out of business of course, who proclaim they are 'gone in.' Men stand at the door of their stores, look up and down the street, yawn, stretch themselves, and retire to mourn over dull times.... Times are dull everywhere; business is stagnated and people want to know the reason of it."[39]

It would be an over-simplification to blame these hard times completely on the state of the hydraulic mining industry. Other factors were also contributing to the dull times in the mining camps. The easily worked shallow placers had "played out" by the mid-'50s and hard rock mining was yet to become a significant factor in gold production. During the decade of the 1850s, this branch of the industry only accounted for one percent of California's yield.[40] Nevertheless, the quartz mines and mills employed a significant number of miners. Between 1850 to 1860 a large number of quartz mines were opened in California, unhappily, by men with no experience or knowledge of the business. As a consequence most became ruinous failures. In 1858, there were 280 quartz mills in California but by 1861, the State Geological Survey found that only 40 or at most 50 were still in operation, and of these several were in danger of closing.[41]

The mass exodus from the mining towns of California was caused by reports of gold and silver strikes in the Fraser River region of British Columbia, the rich Comstock Lode, over the Sierra, and later, the Reese River and White Pine mines of central and eastern Nevada and similar strikes in other parts of the west. In the spring of 1858, when the news of the gold excitement on the Fraser River reached California, the migration commenced and by July, 18,000 men had left for the mines in Canada. Then, a year later, in the spring of 1860, the Washoe excitement became a haven for the disappointed, empty-handed throngs returning from Canada and for a host of unemployed miners who had remained in the idle gold camps of California. John S. Hittell, from a contemporary vantage point, added this perspective to the headlong rush: "The country had become full of men who could no longer earn the wages to which they had become accustomed. Not willing to go down to the valley to farm for $30 per month they had become industrially desperate—demoralized by prosperity."[42]

"The cities and towns in California are losing a large amount of population and considerable money in consequence of these late new and extensive discoveries" declared the *Mining and Scientific Press*. "The rush over the mountains at the present time is without parallel in any former period."[43]

By 1863, the promising young town of North Bloomfield located on the rich auriferous gravel deposits of the San Juan Ridge was nearly depopulated.[44] Small, marginal hydraulic mines throughout the gold country, after 1860, were beginning to disappear and by 1868, one-third to one-half were out of operation either because the claims were too restricted in size or the owners lacked the capital for proper development. The most expensive requirements for a productive and profitable hydraulic mine consisted of an adequate, reasonably priced water supply and a bedrock tunnel, not only to serve as a necessary outlet for mining debris but also as a desirable location for long strings of sluice boxes.[45]

Rampant stock speculations, phony promotions and endless litigation were all an attendant part of the silver bonanza at Virginia City and all contributed to the depressed conditions in California's mining industry. This sentiment was roundly expressed in an article of a leading mining journal.

> The depression in the Mining Share Market, heretofore, of late noticed, still continues. So great indeed, has it become, that it has already resulted in an evident disposition on the part of the capitalists to call in their loans and reduce their risks on this class of stocks. The causes of this depression are not attributable so much to any lack of confidence in the ultimate success of the leading mines as the prospect of endless litigation which seems to hang over them ...
>
> It has now become the especial curse of our land, retarding our prosperity in its most essential point—the development of mines—thereby turning from this coast the flow of capital which would otherwise set thitherward

with a volume and uniformity without parallel in the history of past commercial transactions.[46]

Again, the thoughtful and observant historian, John S. Hittell, lists yet another factor leading to the faltering conditions in the gold fields of California during this period—the federal government's policy of forbidding the sale of mineral lands. As a consequence the inhabitants in the mining regions were unsettled with nothing to tie them to any one place. Thus a large portion of the mining population was nomadic in character. Hittell observed: "Most of them have poor claims, or none at all; and they enact laws, or establish customs having the force of laws, that all claims shall be small usually not more than one hundred feet square. These small claims are worked out in a month or two, or at most in a year or two, and then the miner must go." The transient condition of the miner sired yet another evil—substandard housing. "The dwellings throughout the mines are, as a class, mere hovels, even in the oldest and most thickly-settled districts. In the towns it is necessary to have some substantial stores, as a protection to the valuable goods kept in them; but with these exceptions, and a few fine residences, even nominal 'cities' are collections of shanties, scattered about with little regard to order. . . . The wandering character of the population, and the want of permanent and comfortable homes, render the mines an unsuitable place of residence for families."[47]

The absolute dependency of hydraulic mining on a continual and abundant water supply would always be the source of its greatest vulnerability. The industry's exposure to this axiom was even more critical during the years of its infancy. It would not be until a later period that the great water systems, which would help sustain its mature years, were fully established. But even as late as 1875 the system was not adequate to meet the needs of the industry during prolonged periods of drought. In March of that year a mining town newspaper affirmed the problem. "The water companies are nearly all drawing from their reservoirs to supply the ditches now. We learn that at Iowa Hill in Placer County, the supply will soon give out. In this county [Nevada], the season will be shortened considerably."[48]

During the very earliest years of its development until 1857, the hydraulic mining industry had only one year—1856, of adequate rainfall to allow the new method to reach maximum efficiency.[49] The years, 1862 (after the winter of '61-'62) to 1865, were the worst drought years in California history. From all parts of the gold region could be heard a growing and familiar lament. In March of 1864 The Downieville *Mountain Messenger* reported: "The mines around La Porte have about given up all hope of being able to do anything in hydraulic mining this season. Claims at this place are mostly hydraulic diggings and require large heads of water to run the gravel off and are as rich as any in the country. Most of the mines in the different towns of this county are tunnel claims, and have the advantage over our locality as they do not require as much water to wash their dirt."[50] The prolonged drought was, therefore, a significant contributing cause to the years of depression in the gold districts of California as surely as it wiped out the great cattle herds in other parts of the state and hastened the end of the great rancho era in California.

Dependency on the weather, as any farmer will vouch, can also have a psychological influence on the spirits of those who are subject to its whims—ranging from depression to optimism. The welcomed break in the long drought, beginning in the autumn of 1865, was met by joy and a feeling of renewed confidence by the people in the mining towns. As early as November, the mines in the vicinity of Red Dog had an abundance of water for hydraulic mining, but only a few were ready to commence work. An article in a Grass Valley newspaper exclaimed: "The late storm started all hands and miners are everywhere busy getting ready. In many claims, however, in other parts of the country 'piping' has already commenced and large quantities of gold dust will be brought into market."[51]

By March of the following year it was obvious that the rainy season of 1865-1866 would be protracted. The South Yuba Canal Company reported that from September 1, 1865 to March 21,

1866—44.76 inches of rain had fallen and during that time nearly all of the claims in the county had been worked without interruption. On March 23rd the Nevada *Daily Transcript* expressed the idea that perhaps they were getting too much of a good thing. "It rains, rains, rains and looks as though it would never get done raining." On April 1st the South Yuba Canal Company again reported the seasons rainfall—a total of 51.98 inches.[52]

The depression in the mining industry was not over, but the break in the prolonged drought was certainly a "jump start" to the economy and gave a lift to the spirits of the mining community. It proved a good beginning on the road to recovery. Adding credibility to the reasons for optimism, as early as January 1865, it was reported in a Grass Valley newspaper, that local mining stock sales were booming. During a period of just ten days an aggregate total of nearly $100,000 had been sold. The report concluded: "Quite a business for a small township, but nothing of what it will be in the coming summer."[53]

The following month another report made it clear that the recovery was hardly confined to a local area but was state-wide in extent. "From every point we continue to hear favorable accounts; water being abundant, the miners are fully improving their opportunity, and working with a will, vigor, and success, rarely ever witnessed at any previous season. Very few idlers are found about the mines, and while all hands are busy gathering in the golden harvest, money matters are everywhere becoming sensibly improved and business men once look smiling and hopeful."[54]

Near year's end the enthusiasm had not abated and it was estimated that there would be at least one-third more placer mining that season than the previous.[55] In March of the following year an article in the *Mining and Scientific Press*, after acknowledging that the aggregate yield of the mines had been influenced by the wet season, proceeded to list other factors thought to be contributory to the recovery "such as large numbers engaging in them; the discovery of new diggings; the reopening and working over with greater care, or by improved appliances . . . penetrating further into the gravel deposits; larger crushings from the cement beds, and a general enlarging of the area of the mines."[56]

Old mines which had not operated for four or five years were being reopened. On the San Juan Ridge the Old Star Company owned by the Middle Yuba Canal Company was put back in operation in the spring of 1866 after having laid dormant since 1861. The company was placing its faith in a new four-inch nozzle, at that time the "largest on the coast." It was being operated with 100 feet of pressure using from four to six hundred-inches of water per day.[57]

The same theme of reopening old mines was reported again and again. One such item in a Nevada City newspaper related: "We learn that nearly every claim from North San Juan to Timbuctoo [Yuba County] is being successfully worked. A great number of those claims between those two places have not been worked for years until this season. They are all paying well and employing hundreds of men."[58]

Reports of the mining recovery were not being ignored in San Francisco and an increasing number of investors (capitalists according to contemporary usage) were observed visiting the various mines throughout the state. Nevada County appears to have been the favored destination.[59]

Capitalists were doing more than just looking. In April it was learned that San Francisco investors had recently purchased the Eureka mine located in North San Juan. At that time it was the leading hydraulic mine in that part of the state. After an investment of $180,000 it was then paying from $15,000 to $20,000 at each run of from five to ten days. The largest clean-up, after ten days was $30,000.[60]

In addition to reopening old mines new ground was also developed and large numbers of new claims were filed. This growing trend led the Nevada *Daily Transcript* to report: "The old theory that the gravel ranges of the county were worked out is exploded."[61] Giving credibility to this boast, the Nevada County Recorder's Office showed a listing from January 1, 1866 to October 8, 1866 of 10,358 names recorded. The claims averaged about one hundred feet each, making a total of 1,635,800 feet located in less than ten months. These claims included quartz, gravel, and cement, and also locations for water privileges.[62]

At North Bloomfield, a mining camp that had suffered hard times but was destined later to become a legend, principally through its success generated from the investments of outside capitalists, sounded a prophetic plea in a letter to the Nevada *Transcript*. "The claims here are all good, and could a few capitalists be induced to part of their surplus cash in running a tunnel into this place, instead of their buying up 'wildcat,' not only would they be largely remunerated, but would be the means of enabling us to work our claims the year around, instead of three months each year, and the way we would roll out the gold would astonish everyone."[63]

By 1866, although there had been a steady increase in drift mining, most placer gold was obtained by the hydraulic process. However, at that period piping, for the most part, was still being accomplished through two-inch nozzles attached to hoses. Also, except for the invention of the gooseneck nozzle, and a later innovation designed by C. F. Macy in 1863, resulting in a rifled nozzle, there had been very little improvement in this appliance. It is noteworthy that the four-inch gooseneck nozzle, introduced that same year, gave rise to the term "monitor" with reference to large nozzles. The yield in some claims paid $100 per day per man and in others two or three times that amount. However, the average was probably from $10 to $15 per hand per day. Half of this amount went to wages, water and other expenses.[64]

One of the richest auriferous gravel channels on the Sierra slope was known as the "Blue Lead" which ran through Hunt's Hill, Red Dog, You Bet and Little York in Nevada County then on to Dutch Flat and Gold Hill in Placer County. The term "Blue Lead" was actually a misnomer applied during the 1850s by miners who were under the mistaken impression that a single Tertiary channel with many branches extended from Mariposa to Plumas County. The rich bottom stratum of gravel was usually greatly impacted (cemented) characterized by a deep blue color. The upper, less productive layers, were a reddish hue caused by long exposure of the oxides to the atmosphere. The word "lead" was a corruption of the word "lode" because of its resemblance to an ore-bearing lode or vein.

The channel was three-fourths of a mile to one mile in width and the depth of the pay dirt varied from 80 to 180 feet from the surface to the bed of hard cemented gravel. With the introduction of hydraulic mining, no other area had greater gold production than this channel. The Blue Lead was vigorously washed by the new method until the cemented layer was fully exposed bringing hydraulic operations to a stop except for a section of the channel at Eureka Hill where the cement was sufficiently soft to be worked by piping.[65] The small nozzles and low head pressure were simply not adequate to the task. These limitations were noted by J. Ross Browne with reference to *uncemented* ground after periods of drought. "Where the hills have been dried by constant heat and drought, the clay becomes so hard that the hydraulic stream with all its momentum does not readily dissolve it, and much of the water runs off nearly clear through the sluices, and thus is wasted for the purpose of washing."[66]

The tough cement was formed in Tertiary deposits by the superimposed weight of the debris above it and the chemical action of the intermixtures of sulfides causing a compaction, rock-hard in its consistency.

As early as 1857, it was decided to treat the cement the same as quartz and crush it in a mill. The first cement mill was introduced at Little York by the Massassauga Company that same year. It differed little from a quartz mill but used no screens. The cement was thrown into the battery and crushed by the stamps and then washed into the sluices by a stream of water. Much of the material did not get pulverized and was left with the tailings to be exposed to the weather for a year causing decomposition. The following year it would then be run through the mill a second time. This proved to be a rather inefficient method and was later abandoned.[67]

In 1865 the Massassauga claim was purchased by a new owner

and a modern 20-stamp cement mill was erected. The failure of the previous mill was blamed on the state of the art in 1857.[68] The new mill was a vast improvement and led to a succession of similar plants constructed the following year. In March of 1866, the Nevada *Daily Transcript* ran the following story:

> During the past winter the first cement mill erected upon the Blue Lead north of Greenhorn Creek was built by Hagedorn, Larkin & Company at Hunt's Hill . . . The Blue Cement Lead crosses Greenhorn from Bunker Hill to Hunt's Hill and then runs parallel with the creek. It has been prospected as far north as Jacobs & Sargents Green Mountain Claims. In many places . . . the top has been washed off by hydraulic mining and the lead is for some distance laid bare. At other points it is worked by tunnels. The channel is from one hundred to three hundred feet wide and extends entirely through Little York Township to Gold Run in Placer County.[69]

In the early spring of 1865, even before the new mill at the Massassauga claim was installed, the Grass Valley *Daily Union* noted that new cement mills were operating with great success. "The Blue Lead is absorbing much of the activity of the miners in that region," it remarked, "Though at Chalk Bluff ordinary hydraulic diggings are being extensively worked as is also the case at Buckeye Hill and Quaker Hill."[70]

The new cement mills were designed with stamps of about one-half the weight of their quartz counterparts. One stamp would crush from four to six tons per day. The cement, which was not pulverized as fine as quartz, escaped through a sheet metal screen punched with holes from one-sixteenth to one-eighth of an inch in diameter. The gold was then washed into a sluice where it was separated in the usual manner.[71]

All of the new mills appear to have operated successfully, particularly in Little York Township. In one week's run, it was reported, the Neice & West Mine had cleaned up $6,000, and for six successive weeks its yield was never less than $3,500 per week. A mine near You Bet did even better, cleaning up in ten days nearly $12,000. The mines of the Blue Lead, using these new techniques, proved to be among the richest in the state. The Brown's Hill Claim, for example, between 1865 and 1870 produced over $600,000. And the You Bet Mine, after removing about 50,000,000 cubic yards of dirt, produced a total of nearly $3,000,000.[72]

In May 1866, the Nevada City *Transcript* did a follow-up story on the success of the new cement mills, reporting: "The mills in Little York Township are driving ahead night and day, and all the places in the locality are gaining in wealth and population. The amount that is being taken out continues to be exceedingly rich. One company cleaned up last week $6,380."[73] The following day the same paper informed its readers that Mallory & Company at You Bet struck very rich cement. In fact it was claimed to be better than any bank on the Blue Lead and extended the entire length and depth of the lead. Thousands of dollars in gold were visible and the claim had to be the most extensive and richest in the state.[74]

By 1867 there were 21 cement mills in Nevada County, 16 in Little York Township alone with a total of 136 stamps. During that year it was estimated that about 500 men were employed in the mills and cement mines in Nevada County.[75]

Some credit for the success of the new cement mills was due in part to the introduction of an appliance known as the "Cox Pan" which was designed to separate the worthless boulders and cobbles from the gold-bearing clay. The pan was constructed of iron, six feet in diameter and 18 inches in depth. It was equipped with four iron rakes attached to arms that were bolted to a vertical control shaft. The shaft was geared to complete fifty revolutions per minute setting up an action which separated the clay from the larger material. A steady stream of water flowed into the pan washing the clay through longitudinal openings in the bottom. The rocks, in turn, were ejected through a trap-door also located in the bottom of the pan. The capacity of the appliance was about one ton of material which translated into a production of from 35 to 40 tons in a twelve-hour day. According to authorities, the Cox Pan improved the milling process and was the source of increased earnings for the mills.[76]

During the fledgling years of hydraulic mining, before pipers learned to cut down banks in ledges, it was often necessary for a team of workers to undermine banks with picks. This was a very dangerous process and led to numerous accidents. In 1860 someone got the idea of bringing down banks with a blast of powder. On this occasion, as an experiment, a shaft was sunk at the base of the bank to a depth of 15 or 20 feet and a small chamber excavated at the bottom. Five or six kegs of powder were then placed in the chamber and when detonated, the bank was successfully brought down.

However, the very earliest recorded blast in a cement mine was as early as 1853 when Rogers & Company, at Little York, started mining cement by blasting and then running the loosened cement through their sluices. They would then save their tailings and allow them to weather for a few months causing decomposition, then sluice them again. In some cases the material was run through the sluices six or eight times. But this was hardly a satisfactory solution.[77]

By 1860 a far better method of blasting had evolved. The process involved driving a tunnel into the face of a bank of cement for a distance of about three-fourths the height of the bank and make a cross-cut at right angles in either direction forming a perfect T. The length of the cross-cuts combined would equal the length of the adit. Kegs of black powder would then be brought in and lined up in the drifts which were packed and sealed tight. The ensuing blast would succeed in loosening the ground so that it could be attacked by the nozzles, or in some cases, sent to the mill.

During the early stages of this new process the blast was quite modest. After a trip to California, J. Ross Browne remarked that there had been an instance in which two tons of powder had been used at one blast in a hydraulic claim, suggesting that it was of enormous proportions. Within a very few years explosions of many times this magnitude would be common.

As the amount of powder was increased, the explosion often caused unexpected damage. A blast of three hundred kegs of powder, at the Golden Gate Claim at North San Juan shook the whole town and caused considerable damage to the mine, breaking pipe, causing flumes to come crashing to the ground and smashing sluices.[78]

Blasts of this magnitude became quite regular in their occurrence on the Blue Lead, but were usually worthy of a story by the local newspaper.

> The Gouge Eye Company at Hunt's Hill discharged a blast of fifty kegs of powder. In their claim, the gravel banks above the cement lead pay well for working by the hydraulic process and their object was to loosen this surface ground in order that it may be more easily washed. A tunnel was run sixty feet into the bank and then a drift was made at right angles with the tunnel. In the drift the powder was placed in boxes and as closely confined as possible. The tunnel was then blocked up and a double fuse run to the surface. The blast shook the ground like an earthquake and loosened up thousands of tons of gravel. When this is washed off, the cement channel will be laid bare, and may be easily obtained for crushing in the mill.[79]

From another news account it was learned that an accident had occurred at a cement mine where a fuse had been ignited but the charge had failed to go off. In an attempt to drill it out, the old blast was accidentally discharged causing a miner to loose his eyesight.[80] Similar accidents, unfortunately, would become all the more frequent with the expansion of hard rock mining in the years to come.

The largest blast reported in the state during this early period was conducted by the American Company at Sebastopol using 510 kegs of powder in a drift ninety feet long costing $2,300. The fuse was 45 minutes in burning before it reached the powder. From this single blast it was estimated that the loosened area would yield up to $20,000 for 20 days work.[81]

For many years bank blasting remained a spectator event attracting crowds of viewers. At Mark's & Company's claim at Brandy City, Sierra County, a blast containing 450 kegs of powder, the largest ever by 50 kegs in that camp, prompted the *Mountain Messenger* to report: "At 5 o'clock in the afternoon the banks opposite the mine were lined with women and children to witness the scene." [82]

In addition to upgraded methods of cement mining, by the late

'60s, other changes were taking place in the industry including: the excavation of drainage tunnels; the consolidation of claims; increased outside capital investment; and the construction of new water ditches—all contributing to improved conditions in the industry.

In late spring of 1866, with the completion of the Gold Run Ditch in Placer County, the mines in the vicinity of Scotts Flat were allowed to commence work after having been shut down for several years. Stories of the revival of mining as the result of the extension of old ditches and the construction of new ones, were becoming common throughout the mining region. Upon the completion of the Omega Ditch in Washington Township the *Transcript* noted, " . . . there are some nine or ten companies at work and everyone is doing well. The diggings are all hydraulic and are supplied with water from the new Omega Ditch, which will now enable the miners to work the year around."[83]

At the mine sites, drainage tunnels, which were a necessity in most well-engineered hydraulic operations, were generally driven from the most convenient location on the claim and designed to approach near the center of the gravel deposit on a grade of four to five percent. Tunnels could vary in length from a few hundred feet to over a mile and were generally six or seven feet in width and seven feet in height. From the bottom of the gravel deposit a deep shaft was customarily excavated through bedrock before driving the tunnel, which was begun at the outlet end. In this manner the proper level was obtained avoiding the danger of driving too high and coming out above bedrock. Most tunnels were driven a little to one side of the vertical shaft with a cross-cut made to connect the bottom of the shaft with the tunnel. Vertical shafts would often have a depth of 200 feet.[84]

Bedrock tunnels were sometimes excavated by the mine owner using his own employees, however, most work was done by contract using sealed proposals or bids. Prices ranged from $34 to $37 per lineal foot when driving through hard rock. Each tunnel worked from six to nine men per day with two shifts. If shafts were driven, two faces could be worked at the same time. In some cases it was necessary to drive a second tunnel, at a later date, to reach a lower level. It was felt that if the first tunnel had been initially excavated at such a depth, the bank would become too high to work safely and efficiently. In later years, with greatly improved monitors, this would not be necessary.[85]

But even including this expense, according to one authority, unless the bedrock tunnel was extremely long, it was still cheaper to open a hydraulic mine than a drift mine. It was estimated that the cost of drift mines ranged between $100,000 and $200,000, while hydraulic mines were considerably less.[86]

One of the earliest completed tunnels was the Yuba Tunnel at Sebastopol in Bridgeport Township, Nevada County. During the six years of its construction beginning in 1860, all operations at the mine had ceased because it lacked a proper outlet for its tailings. However, by 1866, the mine was hailed as a model for other operators to emulate for it was changed from a piece of unprofitable ground to one of the richest mines in the state.[87]

The following month yet another success story was reported by a local newspaper after the completion of a tunnel. "The American and Knickerbocker Company have one of the best tunnels in the country. It cost between $25,000 and $30,000 and they are now reaping the reward of the enterprise. Before the tunnel was run the claims were valueless, it being impossible to work them to run off the tailings."[88]

In some cases, if the owner had a little capital and a lot of foresight, through intense labor it was possible to open a rich mine without the aid of outside capital. The Blue Gravel Claim at Smartsville in Yuba County was opened as early as 1853 with a bedrock tunnel begun the same year. During the first nine years the mine produced $315,489 which was all reinvested to pay current expenses plus an additional $7,543 supplied by the owner-operators. The tunnel was completed in March 1864. During the following 43 months the mine produced $837,409 of which $625,543 were net earnings. The average monthly expense was less than $5,000.[89]

The efficacy of bedrock tunnels as a means of increasing the productivity of hydraulic mining ground was a frequent topic in the Nevada *Transcript*.

> A survey has been recently made for a tunnel to run from the Yuba River at Chimney Hill which is situated about half a mile from Cherokee in this county. The hydraulic gravel claims in that vicinity, which are very extensive, for want of an outlet, can now be only worked to a depth of from 40 to 100 feet. When the prospect tunnel is completed, the same ground can be worked to a depth of 300 feet and over two miles of surface will be opened which is now impossible to develop.[90]

The excavation of bedrock tunnels became extensively adopted throughout the hydraulic mining districts not only to revitalize old mines but as a necessary adjunct to the opening of new ground. Many would be driven for thousands of feet. By 1870 a tunnel completed by the Welch Company near French Corral, 2,500 feet in length was at that time considered the longest in Nevada County. Two years later, the Bed Rock Tunnel Company drove a excavation 2,900 feet through a hill near Birchville with a connecting shaft 249 feet deep—the longest in the state. A short time later The Gold Run Ditch & Mining Company of Placer County ran an eight-foot by ten-foot bedrock tunnel 4,000 feet in length, while nearby, the Cedar Creek Company at Dutch Flat ran their tunnel a similar length. A number of mines had tunnels from two to three thousand feet, including the Nevada Mine at You Bet and a number around Smartsville in Yuba County. The North Bloomfield Gravel Mining Company exceeded all others with a tunnel at the Malakoff Mine—just short of 8,000 feet.[91]

Working conditions in some of the longer bedrock tunnels often became hazardous because of poor ventilation leading to foul air. To alleviate this problem a device called "the water blast" was invented. It was a very ingenious contrivance in which a perpendicular fall of water forced air through a four-inch pipe to the working location of the miner.

With the passing of the late 1860s, hydraulic mining was emerging from its adolescence toward maturity. But it would not be until the dawn of the '70s that the great nozzles, destined to be called giants, would reach fruition, impelling the growth of a water system without parallel anywhere else in the world and bringing the industry to its fullest potential.

CHAPTER 3

Water, Ditches and Dams

As the decade of the 1860s neared its end, hydraulic mining was experiencing a strong revival. The previous years of boom and bust had led to a better understanding of the requirements and true dimensions of the new industry. During that rather hectic period the miners had learned to deal with cemented layers of gravel; had discovered the need for bedrock tunnels and an adequate and reliable water supply; had learned to consolidate their limited claims; and finally, the miners had become aware of the growing necessity for venture capital to implement long-range development.

Interest in hydraulic mining as an investment opportunity was no longer confined to San Francisco capitalists. Although most western mining was then shunned by New York money interests because of their unhappy experience with the wild speculations and stock manipulations on the Comstock, it, nevertheless, attracted considerable attention from investors on the other side of the Atlantic. In the early spring of 1871 the Birdseye Creek Gold Mining Company, with headquarters in London, was among the first to take the risk by purchasing extensive mining properties in the vicinity of Little York and You Bet.[1] A year later British enthusiasm for hydraulic properties had not abated. "A number of English capitalists arrived in this city on Wednesday evening last and are staying at the National Exchange Hotel," the Nevada City *Transcript* reported. "Professor Silliman is also here, together with other prominent men of San Francisco. Our mines are not only attracting the attention of San

Francisco but English capitalists. The object of the latter to our city is for the purpose of negotiating for the purchase of the best placer mines on this coast, those of Sargent & Jacobs at Quaker Hill, and several claims at Blue Tent."[2] On another occasion the paper noted that the profitability of the mines was at last becoming known. "The mines of Nevada County are beginning to attract among capitalists, the attention their merits have long deserved. We are informed by hotel men here, that there have at least 30 mining experts and capitalists, wishing to invest in mines, arriving here each week, for the last month or more. They come up to look over the field, and see for themselves."[3]

It should be noted that the California hydraulic mines did not always remain unattractive to New York money. Toward the end of the '70s, after establishing a profitable and stable track record, the attitude of eastern investors began to change. A September 1880 item in the Downieville *Mountain Messinger* is indicative of this trend: "Arrangements are under way to sell the American Hydraulic Company's diggings at Morristown to New York capitalists. $165,000 was taken out of this claim in one year. The gross yield for the past 28 years since 1852 has been over $7,000,000."[4]

New ditch ventures, likewise, were attracting outside investors, but not without assurances of sound business practices and positive profits. This branch of the mining industry had passed through a very perilous learning period. By 1870, although over 6,000 miles of ditches had been constructed at a cost in excess of $15 million, and while many were profitable, others had been ruinous failures.

Early California ditches were often run with grades engineered too steep causing rapid erosion resulting in frequent breaks and large maintenance costs. The other extreme could prove equally difficult. If the grade were too slight it often caused excessive silting and eventual stoppage. Such a disaster became a reality for the Eureka Lake Company when, in the Spring of 1866, their main ditch was clogged with sediment all the way from North Bloomfield to Eureka South (Graniteville). As a consequence, the hydraulic operations in that rather vast area were shut down during a very busy season.[5]

A variety of other problems also plagued the ditch companies. Upon one unfortunate occasion the hydraulic mines in the vicinity of French Corral were forced to stop operations in mid-season while the company cleared a large slide near Badger Hill caused from building on an unstable bank.[6] From such experiences it was learned that a well engineered ditch needed to be designed in such a manner as to keep the velocity of the water, as well as the greater force of the friction back against the hill as much as possible, however if this force were excessive it could cause extensive undercutting and possible slides. Ditch builders also learned that ground which was apt to give way should be flumed in spite of the temptation to dig the ditch in soft ground. It became axiomatic in the hydraulic mines that nothing was more costly than shutting off water to repair damaged ditches during the mining season and thus preventing the monitors from continually washing gravel into the sluices.[7]

Even well-established ditch companies had their share of problems before learning to take the precautions which would later become basic operating procedures. In localities where it was determined flumes were subject to excessive decay, it was found that stringers should be laid on the ground under the sills the entire length of the flume. In July 1868 the flume of the South Yuba Canal Company gave way in the vicinity of South Steep Hollow from such an oversight causing the excessive decay of timbers which were unable to withstand the weight of the water. Some eight sections were torn away from the bank, but it was estimated that at least twice that number would have to be removed and replaced to restore the flume to a safe working condition. In an effort to get the repairs done as quickly as possible, a large gang of men were put to work, laboring day and night to hasten the completion date.[8]

But these problems were rather minor compared to a number of engineering fiascoes caused by the *naiveté* of uninformed investors in the mid-'50s. A Sacramento newspaper, in a rather derisive manner, expressed the frustration of the mining community:

The few efforts which capitalists made in attempting to supply mining districts with water proved quite abortive. They relied upon *their* knowledge and disdained the advice of the red-shirted miner. In the County of Sacramento one company of San Francisco capitalists sunk $150,000 in the vain attempt to supply Mississippi Bar with water for mining uses by means of steam engines and force pumps, and another company wasted $75,000 in the same way on Texas Hill. The engines employed in the latter operation are now used in the water works in this city. If the parties who had engaged in these affairs had known what amount of water was required by a miner for a sluice, and what he could afford to pay for it, they never would have attempted to supply the article by steam power. At the same time a company of miners in the same vicinity were laughed at for attempting to supply water by means of a canal. They succeeded, however, with a capital of $80,000 in bringing a quantity at least fifty times greater than the amount supplied by both the steam engines.[9]

Outside investors were prone to other mistakes as well. Many had no clear perception of the various difficulties that could be encountered during the construction of a project of such magnitude. They also had an erroneous idea of the time required to realize a return on their investment. In many cases money would be borrowed, at very high rates of interest, for the unrealistic period of perhaps six months or a year. When the notes became due, new and larger loans became necessary to satisfy the demands of their creditors. The result of this short-sightedness often caused the forced sale of many ditches at a figure of from one-fourth to one-half the actual value of the property. Investors learned from these failures that at least two if not three years should be allowed before expecting any monetary return.

In 1853 the Miners' Cosumnes & Deer Creek Canal Company began the excavation of a major ditch near Mormon Island, Sacramento County. The main channel was 14 miles long with two lateral branches, one seven miles in length, extending to Walls Diggings. The other branch took its water from the main ditch and terminated at the American River a distance of five miles. Because of the lack of adequate financing, the construction was never completed and, due to its unfinished condition, suffered severe damage during the winter rains of 1854. Like so many others, this underfunded ditch company fell prey to the auctioneers hammer. This incident prompted the Sacramento *Daily Union* to lament: "The depreciated value of the stock of almost every water company in the State, with failure and utter worthlessness of so many, commenced and prosecuted under favorable circumstances and sanguine expectations, present to miners, but more especially to capitalists, a problem of the highest importance."[10]

Unbridled greed extracted a toll from at least one ditch company. The Tuolumne County Water Company, with an investment of $275,000, supplied mostly by D. O. Mills & Company of Sacramento, was the first to bring water to the busy diggings at Columbia, a place distinguished, for a time, as the richest placer mining camp in the state. But the company, which held a virtual monopoly of the water supply in the region, proceeded to charge exorbitant prices for its water. The miners, after a meeting at Farnsworth's Saloon, protested the excessive rates in a written petition, but were unsuccessful. Finally, in utter frustration, they resorted to an all-out boycott of the company. During this period feelings ran so high that some of the miners actually resorted to vandalism, destroying a portion of the company's flumes. This rash action prompted the *Mining and Scientific Press*, in a conciliatory editorial, to urge an agreeable conclusion to the boycott.

> There is a prospect for a settlement of the troubles with regard the ditches of the Tuolumne County Water Company. The company proposes to supply water to the miners at sixteen-and-two-thirds cents per inch (a reduction of one-third the former price), provided they are guaranteed in the peaceful running of water. It is proposed to call the miners together and lay the matter before them . . . we earnestly hope that the ditches will not be destroyed again. The people of Tuolumne County have already suffered immense pecuniary loss, and as long as the present uncertainty exists with regard to the continued supply of water so long will the losses continue. Business is almost paralyzed in the vicinity of the troubles and better times are earnestly hoped for."[11]

Although the boycott was eventually settled, the animosity generated from this bitter experience ultimately led to the construction of a competing ditch, known as the Columbia & Stanislaus Water Company. During the boycott, many of the otherwise idle miners, actually contributed to the construction of the ditch by donating their labor in exchange for shares in the new company. Thus the Tuolumne County Water Company lost its big chance—the record-breaking days of placer mining in the region were soon over, leading to a bleak future for both ditch companies.[12]

The long recession in the mining industry was the final blow for many ditch companies. After mines and whole towns were shut down and deserted, many ditches were sold at great losses or, in some cases, simply abandoned.

J. Ross Browne writing in 1866, in his *Report* as Special Commissioner for the Collection of Mining Statistics, made this pessimistic assessment.

> The ditches . . . of which there were at one time four hundred in the state, owning a total length of six thousand miles of canals and flumes . . . Unfortunately these mining canals, though more numerous, more extensive, and bolder in design than the aqueducts of Rome, were less durable, and some of them have been abandoned and allowed to go to ruin, so that scarcely a trace of their existence remains, save in the heaps of gravel from which the clay and loam were washed in the search for gold.
>
> As the placers in many districts were gradually exhausted, the demand for water and the profits of the ditch companies decreased; and the more expensive flumes, when blown down by severe storms, carried away by floods, or destroyed by the decay of the wood, were not repaired.[13]

With the revival of hydraulic mining and the enormous demand for water and more water and with a better understanding of the vagaries of the industry, even greater investment opportunities abounded for ditch companies—leading to massive consolidations and expansion. Historian Samuel Bowles was thoroughly impressed by the magnitude of these operations recording: "Near Dutch Flat where extensive hydraulic mining is in progress a water company taps lakes twelve to 20 miles off in the mountains and turns whole rivers into its ditches; and as further illustration of its majestic operations, we learned that it spent eighty thousand dollars in one year building a new ditch and yet made and divided one hundred and twenty thousand dollars in additional profits that same year."[14]

To the modern reader, it seems incredible that an individual or group of individuals, could have the right to "tap lakes and turn whole rivers into ditches" solely for their personal use or profit. Yet, from the time of the Gold Rush until well into the '80s, the laws of California, although peculiar were quite specific with respect to the rights of the miner to control the mining ground and water. Although he had no ownership in fee, he had absolute control so long as he conformed to the mining codes of his district.

The rights of a miner to special privileges had its origin in the recognition that minerals and metals have a unique value to organized society and date back to the 12th Century. At Freiberg in Saxony the term *Bergbaufreiheit* came into usage signifying the right of free mining. In later years, in the 17th Century, a similar principle was adopted in the tin mines of Cornwall when the policy of "bounds" was instituted and miner's courts (Stannary courts) were established. These antecedents made an inevitable reappearance in the California gold fields and were carried to extents never before known.

From the time of its very creation under military government, the great business of California was mining and the right of miners were very carefully guarded and esteemed above all other property rights of real estate. With respect to ownership and control of water, the law in the Golden State protected usage unknown in English Common Law. This included the permanent diversion of water from its original channel for mining uses and the right of locating streams for such purposes.[15]

In California, miners could hold property by "usufruct" in unnavigable streams. This strange word is a Latin term used by lawyers to express the right to utilize and enjoy the profits and advantages of something belonging to another so long as it is not damaged or altered in any way. In one case this doctrine was stretched to the point that a mining claim was actually granted a patent in the bed of the Bear River. Charles Shinn, one of the best authorities on this subject, explained it this way: "The gold-seekers could and often did sluice away roads, or cut them across by channels impassable for years, undermine houses, wash away fertile land, move towns to new sites and tear the old location down to bedrock with torrents of water."[16] But as H. P. Davis recalled, at least in one case, there were limits to this callous disregard for community property the townfolks were willing to endure: "At Main and Commercial Streets [in Nevada City], opposite the present post office, two miners started to dig a gravel pit and when expostulated with by indignant merchants, calmly stated that there was no law against mining in the streets. This oversight by the lawmakers was promptly corrected by property owners with shotguns."[17]

But generally Nevada City, like all mining towns, was quite tolerant with respect to any inconveniences caused by the mines. The town's very existence depended on that industry and in most cases the diggings were there first. Even as late as 1874, the sluices carrying sand and gravel from the Manzanita Mine actually ran under the Union Hotel in the heart of town. In June of that year it was reported that the fall was not sufficient to carry off the tailings resulting in an overflow backing up the slickens through the city streets. To rectify the problem, at least on two occasions, the sluice had been lengthened to lead off the tailings. However, the very next day the local newspaper without rancor or complaint reported: "The flume running under Pine Street and from thence across to the Broad Street Bridge, became clogged at its mouth yesterday, and Marshall Gray and his assistants were engaged in cleaning it out. Flumes seem to be a little contrary just now."[18]

Two years later the townfolks at Moore's Flat calmly acquiesced to an even greater inconvenience as reported by the *Transcript*: "The citizens of this place are making preparations to move the town, and work will commence tearing down the houses after the Fourth. The town will be built about a quarter of a mile from its present location. The caving of the surface of the hydraulic mine has caused the necessity of moving. It will be considerable expense, but the citizens have got pluck and courage and will soon have the new town in full blast."[19]

The leniency of the law favoring the miner had its genesis in California, before it was granted statehood while under the military government. For this reason the federal mining laws did not apply in the territory. As a result, the local mining codes for each district, which were already in place, were officially recognized. Even Josiah Royce, who was quite critical of the miner's courts with respect to criminal law, admitted that most contemporary accounts of their actions, in matters dealing with mining law, "they regard as furnishing the only just and truly legal method of dealing with mining rights. They resist strenuously any legislative interference with their local self-government in these matters. They insist absolutely upon the autonomy of the miners' district, as regards the land; and for years against all legislative schemes at home, and all congressional propositions at Washington, they actually maintained this autonomy."[20]

While constructing a dam on the South Yuba River near Irishman's Bar, Peter Decker recorded in his diary an example of the efficacy of the local mining codes and miner's courts in dealing with matters involving water rights.

> Thursday Aug 1st [1850] . . . Mull & Co. Who have river claims above ours notified our men today noon of our backing water on them & for us to stop work which we did for the afternoon, & committee from both companys met to settle but failed. They listened to our companions reasonable & our work will now be forwarded & the *priority* of rights of locations of claims will have to be tested.
>
> Friday 2 . . . Most of the day the dam hands were engaged diging the head of the race deeper to back up the water less on Mull & Co. And enable us the better to turn it in.
>
> Saturday 3 . . . Committee from Mull & our company met again today, to settle the dificulty of our backing water on them but failed again to settle. Mull & Co will call the miners together on Wednesday at 1 O'clock to have the matter decided by a jury—hearing evidence etc. Our dam backs the water through their whole race. . . .
>
> Wednesday 7 . . . Our Company and Mull Loree & Co. Had a trial this afternoon as to priority of claims caused by our dam backing water on their

claims. 100 miners were present—Mr Lyons of Iowa was lawyer — Mr Slater council for plaintiffs. Pease Moderator or Judge — rules as in court. Had 12 jurymen examined some 20 witnesses, agreed in written rules that a majority of jury should decide or when evenly divided the judge to give casting vote. Which being the case the judge decided in our favor. Trial orderly—formal and notwithstanding the parties bound themselves to abide by the decision, yet Mull & Co much exasperated.[21]

The right of discovery and appropriation of mineral property and priority of water rights, as a source of title, became the law of the land. The federal government owned nearly all of the mineral land in California and permitted the miners to occupy the land for the purpose of mining with the express consent of the state. The federal government also refused to sell the land or open it for preemption for agricultural purposes. The only exception to this policy was for mineral land that had been a former Mexican land grant as in the case of John C. Fremont and the Mariposa grant. In short, any man had the right to take up a claim by going to the County Recorder pay a fee and give a description of his claim.[22]

An excellent account of the construction of one of these primitive ditches for the purpose of bringing water to a hydraulic claim at Chamber's Bar, in Tuolumne County, was recorded by Howard C. Gardiner in the latter months of 1855.

> I devised a simple instrument to mark the line of survey. This was nothing more than a straight-edge strip six feet long, to which was attached at right angles another strip five feet long. A plumb bob in the center of the straight-edge fell along the upright leg, which had a hole for the bob to rest when the leg stood perpendicular, and when it did so, the straight-edge was of course perfectly level. A pole five feet one inch long was prepared and sights were taken every two rods. When this pole stood on the ground so that a sight along the level straight-edge struck its top, its bottom was just one inch lower than the bottom of the leg of the straight-edge; a stake was driven to mark the spot . . . this gave the ditch a fall of one half an inch to the rod. This fall afforded a good current and enabled the conduit, which was two-and-one-half feet at top, eighteen inches at bottom, and twenty inches deep, to carry an ample supply of water. The flumes, which were made of sluice lumber, were given a good deal greater fall than the ditch, and though smaller had a greater capacity . . .

> When the survey was completed and verified, we proceeded to contract with some Chinese for grading and excavation of the water course. The Celestials were paid one dollar and a quarter per rod for construction.... The whole length of the ditch was three-and-one-half miles....
>
> Some thirty Chinamen were engaged on the ditches.... The gulch, which in times of rains was a raging torrent, required a substantial substructure for the flume, and when the trestle work was completed it was really a creditable piece of engineering. The conduit across the ravine had a fall of twelve inches in its length of 120 feet....
>
> The flume in the center was 100 feet above the bottom of the gorge. I was my own engineer, cut the trees, and placed the timbers, with the assistance of one man and a horse.[23]

Any one man had the right to take possession of a whole stream and direct its waters to mining ground and charge any price he felt the traffic would bear. At first the miners objected to the idea of companies acquiring land solely for the purpose of selling water to the miners and not for their own use. As a consequence a number of lawsuits were filed resulting in a state supreme court ruling affirming that the retailing of water was permissible under state law.[24]

Historically water had always been considered a possession of the community to be used for the common good. This doctrine was respected by both primitive and civilized cultures, but as Norris Hundley related: "[The] attitude toward water changed after the discovery of gold, when thousands poured into California imbued with a spirited individualism and an appetite for profit that elevated the exploitation of nature to new heights...."[25]

All that was required to establish a right-of-way for a ditch was to survey its course, plant stakes and file public notice of intent to construct the ditch.[26] For example, in 1852, Carson Wilkinson & Company filed the following notice with the Sierra County recorder at Downieville, "... we have taken up and claim the Water Privileges of the North branch of the Middle Yuba & of Wolf Creek together with the tributaries for mining purposes, and that we intend to conduct the waters of said streams to Minesota [sic] diggins for the purpose aforesaid as soon as practicable." The following December, W. G. Still, County Surveyor for Sierra County filed this follow-up: "I hereby certify that I have this day completed the survey of the route for a ditch, for the purpose of bringing the waters of the north branch of the Middle Yuba & Wolf Creek & their tributaries to Minesota that I gave a fall of ten feet to the mile & that a fall of twenty-five feet to the mile can be given if required. The whole length of the ditch will be eleven miles ninty four [sic] rods as surveyed by request of Carson Wilkinson & Co."[27]

With ditches crisscrossing the countryside and many finding their source in the same stream or tributary as another, it is little wonder that there ensued a myriad of lawsuits over water and mineral rights. Discrepancies in compliance with the law or some informality whereby a shadow of a right was given for someone to either "jump" a claim or institute a lawsuit over water privileges became common. The first lawsuit in Nevada County over water rights had its origin in November 1850, when the Deer Creek Water Company began a ditch at the upper end of the creek while the Coyote Water Company began another, in the same creek, next to town. Upon completion, both companies became involved in continual lawsuits over priority of rights. The matter was finally settled by a consolidation in 1851. Another similar suit was filed by the Nevada County and Sacramento Canal Company in District Court against the South Yuba Canal Company in 1855 and remained in litigation until 1872. During that time many preliminary issues had been tried and the case had been to the Supreme Court several times. This case ended, like so many others, with the purchase of the smaller company by the larger. The South Yuba Canal Company was a veteran of such actions beginning with a legal dispute involving the Coyote & Deer Creek Water Company. Litigation over water rights began in 1853 finally ending with another of many acquisitions, consolidations and buy-outs which would eventually give the South Yuba Canal Company the most extensive and valuable water rights in the county.[28]

During the construction of ditches it was a general practice to install waste gates at appropriate intervals along the extent of the system for the purpose of allowing the water to escape after reaching a certain level. In this manner the banks of the ditch were protected from undue pressure and the danger of overflowing or being washed away. Ditches could vary from two to ten feet wide with a flat bottom and sloping sides designed to insure stability and avoid caving. However, if the ditch was being driven through rock, vertical sides were permissible. For the best results, the width was kept to about twice the depth. A well-designed ditch would not run water more than three-quarters full, allowing from one to three feet of freeboard. The grade was usually engineered for a fall of between four to eight feet per mile or three-quarters to one and one-half feet per one thousand feet. As mentioned before, the early ditches were often run with much steeper grades resulting in too rapid erosion and frequent breaks. The degree of grade had to be designed in relationship to the type of soil it was passing through, ranging from loose sand or gravel to hard rock. It was extremely important to engineer the ditch to retain the exact grade, as any deviation could retard the flow.[29]

The flow of water was measured by means of an instrument known as a weir. It consisted of a wooden dam across the stream or ditch with a square notch in the center through which the water passed. The width of the notch was at least six times the depth of the water flowing over the crest. A stake was driven into the bottom of the pond, above the weir, five or six feet with the top of the stake exactly level with the notch. The depth of the flow over the weir was measured by means of a rule or square placed on top of the stake. The flow could then be calculated by an algebraic formula.[30]

The velocity of a stream was determined by laying off 100 feet along the bank and throwing a float into the stream a short distance above the mark. The velocity was then calculated by the amount of time required for the float to reach the one hundred foot mark. This process was repeated a number of times and an average determined. A properly engineered ditch should not have allowed the velocity to be less than one-foot per second nor more than three feet per second.[31]

All of these precautions were also necessary to guard against seepage. A medium size ditch, five miles long carrying from 1,000 to 2,000 miner's inches could lose from five to ten percent of the intake water by seepage in good soil and in porous soil as much as 20 percent. Some of the remedies to correct excessive seepage were to reduce the size of the ditch, to slow the velocity to the point where the silt would deposit a seal over the ground, to line the channel with canvas or concrete or to construct a flume.[32] The loss of water by leakage could cost thousands of dollars in net profits for a large ditch company. It was a matter of great concern, for example, to the Milton Mining & Water Company as evidenced from a letter written by General Manager Henry Pichoir, from corporate headquarters in San Francisco to his Superintendent V. G. Bell in French Corral: "We are pleased to see that you are giving your personal attention to the matter of leakage of your ditches and we hope that you will not let up until you have very materially reduced the evil. Even if fluming is considered necessary in several places."[33]

Experienced ditch builders had learned from trial and error that when excavating through certain types of soil special precautions were necessary to insure against cracking from the hot rays of the sun. It was considered good practice to allow a small amount of water to be let into the ditch during construction. At Chips' Flat in Sierra County after a hydraulic mining company completed a four mile ditch and turned on the water supply, none reached its destination. It was soon learned there had been extensive cracking on both the sides and bottom of the ditch from prolonged exposure to the sun. The condition was remedied by stationing workmen along the course of the ditch engaged in a process of "puddling" which involved slowing down the flow and causing the accumulated silt to actually seal the cracks. Critics agreed that if a small amount of water had been let into the ditch during construction, the cracking would never have occurred.[34]

While formulating plans for any considerable sized ditch the first consideration of the ditch company was the erection of a portable sawmill to supply the huge quantity of lumber required for the construction of flumes and their maintenance and to provide for the crib work used in the construction of dams. It was considered good practice to locate the sawmill close to the ditch for the purpose of a water supply to power the mill. Another consideration in determining its location was the availability of a gully running from the near proximity up a hill to the thickest timber for purposes of a suitable skid-way. The logs were thus cut to the required length and shot down the skid-way and into a pond or against an embankment.[35]

Flumes were not only necessary where the ditch passed over porous or shattered ground, but also around steep hillsides, banks, cliffs and ravines. As a general rule flumes were constructed so that the width was twice the depth of the water, leaving a freeboard of from one to two feet. They were usually built in 12 or 16-foot sections of one and one-half to two-inch lumber 12 to 24 inches wide. The joints were made tight by means of battens one-half of an inch thick by three or four inches wide. When a flume was constructed on a trestle, a walkway was provided by either a line of planks nailed over the cap or over alternate sills extended a couple of feet from the side of the box.

The average life of a flume was only eight or ten years as they were constantly exposed to winds, fire, heavy snows and rotting. The inspection, maintenance and replacement of worn or damaged flumes was an ongoing cost of doing business for a ditch company and was factored into the price charged for water. Some early flumes were very lofty, 150 feet or even 200 feet high or more. When flumes were constructed along the side of a cliff, brackets were used to support the framework. At the Miocene Ditch in Butte County, William H. Bellows, superintendent of construction, rather than construct a flume on a trestle nearly 200 feet high, chose to have it attached to a vertical cliff 350 feet above the canyon floor. Workmen were slung in ropes over the precipice and dropped down more than 200 feet where they drilled two lines of holes into the wall, one for brackets and the other for supporting braces or suspenders. The brackets were made from "T" rails, projecting ten feet horizontally from the wall to support the flume. These, in turn, were connected to the supporting braces made of heavy round iron. During the construction of the Blue Tent Mining Company ditch, which ran six miles along the face of a cliff, workmen had to be suspended by ropes a thousand feet above the bottom of the gorge.[36]

One of the greatest engineering accomplishments near the close of the Gold Rush era, was the construction of the Big Gap Flume to divert water from the South Fork of the Tuolumne River to Big Oak Flat in Tuolumne County. Completed in 1859, it had a height of approximately 280 feet above the ground, a total span of 2,200 feet, and a center to center distance between each tower of 200 feet. This imposing structure was supported by two wire cables, three inches in diameter with secondary cables composed of ten parallel strands of iron wire with a diameter of one-eighth inch. The concrete foundations of the two largest, of eleven towers, were nearly ten feet thick while the others were 50 feet square. Each section of the flume between towers weighed in excess of 25 tons. This magnificent structure was designed by G. W. Holt and constructed at a cost of $80,000 by contractors Holt & Conrad for the Golden Rock Water Company. Critics, at the time, expressed the opinion that the structure would last at most, two years. However, it served the miners at Big Oak Flat for a period of nine years which was actually the average life-span for ordinary wooden flumes during that period.[37]

In the event of an accident involving a flume, a waste gate, installed every half mile, would be opened to stop the flow of water until the gate at the head of the flume could be closed so repairs could be made. They were designed in such a fashion that the discharged water flowed clear of the line to prevent any possible undermining. Flume tenders were stationed along its course every five or six miles on the alert for any problems that could arise. If, for any reason, the water should fall below a certain level, a float arrangement to which a rope was attached was led through a pulley in the stationhouse where, using pots and pans on the other end, it would set up a racket sounding an alarm to alert the ditch tender.

At higher elevations it was often the practice to install snow guards to cover a flume or ditch along steep banks for protection against heavy snows which were apt to cover the channel to a great depth. These snow guards or snow sheds, as they were also known, consisted of sets of timber placed at intervals of four feet and cov-

WATER, DITCHES, AND DAMS

Flume spanning a valley.
From Taliesin Evans, *Hydraulic Mining in Califonria*.

Method of hanging a flume to a cliff by iron brackets.
From August J. Bowie, Jr. *Hydraulic Mining in California*

Bracket flume of the Miocene Mining Company's ditch, Butte County.

ered with boards or lagging. If the flume was set in close to the bank, the circulation of air around it, during the winter, would be partially blocked by the snow, and freezing was not as likely to occur as when the flume was exposed on all sides. In some cases, even iron pipes were protected from being crushed by the weight of the snow. During freezing weather it was not unusual for ice to build up on the sides of a flume known as "anchor ice." The huge weight from this condition could take out large sections of a flume. With the formation of anchor ice on the inside-bottom of the flume, it was immediately necessary to turn out the water at the waste gate, otherwise the flume would fill up solidly with ice and remain so until spring. For added protection, expert skiers were sometimes used to tend the ditches.[38]

The work of the ditch tender under these conditions could sometimes become very hazardous. At a particularly treacherous stretch of the Milton ditch, between Moore's Flat and Graniteville in Nevada County, during the early spring of 1883, a ditch tender on snowshoes, in the course of his duties, fell into the icy waters of the swift flowing current of 3,000 miner's inches and was swept along into a 300-foot covered section of flume. In the darkness he managed to extricate himself from his snowshoes by actually removing his boots. Upon reaching daylight, after traveling a total distance over one-half of a mile, he managed to crawl out of the ditch, in a freezing condition, more dead than alive, and stagger barefoot to the house of a rancher, nearby. Here he eventually received a new pair of boots and a change of clothes, and reported back to work the following day.[39]

Yet with all these precautions, during periods of very heavy snow, nothing could be done to adequately protect the system. In the early spring of 1874, during near-blizzard conditions, said to have been the worst since the construction of the Central Pacific Railroad, all the ditches of the South Yuba Canal Company were so clogged that no mines could be furnished with water.[40] Again, two years later a Nevada City newspaper reported that all the mines on the San Juan Ridge were closed after the March storms. The article added: "The manner the ditch is opened most of the way, is by working a small head of water from one waste gate to another, until a small opening is secured, then more is let on, until a tunnel is cut through which will allow a full head to run. There is now over ten miles of such a tunnel on the Milton ditch." It was also reported that snow at the head of the ditch was from 15 to 20 feet deep and the Milton Company was working 40 men to get the water moving.[41]

During that same winter of 1876, one ditch that came through unscathed was owned by the Blue Tent Consolidated Mining & Water Company. Thirty miles in length, it had been completed the previous October at a cost of $160,000 with a crew of 1,000 Chinese and white workers. Reporting the day following the story of the snow problems with the Milton ditch, the *Transcript* related: "The ditch around the hillsides, where there is a danger of snow slides, was blasted into rock most of the way, so that during the past severe winter not a break has thus far been made. This method was more expensive, but it will prove most economical in the end. At one of these points 450 rods [7,425 feet] of ditch cost $38,000 for construction. The ditch will carry 5,000 inches of water."[42]

But even the Blue Tent ditch was not immune from freezing during an intensely bitter winter as later reported: "The recent cold weather has caused the suspension of washing in all principal hydraulic mines of the county, including the Blue Tent, Sailor Flat, North Bloomfield, Milton, etc. Water was turned out from most of the ditches as soon as the mercury commenced to go down. . . . Even the large Smartsville ditch [much lower elevation] never before shut down, is only open for a few inches along the center. The water that rushes down is thick with broken ice and snow."[43]

In some cases, particularly in the later period, flumes were often constructed of sheet iron and were protected from rusting by a coating of coal tar.

From the foregoing it is obvious that unless the channel had to be directed through slate or porous ground or along the side of a precipice, ditches were far less trouble than flumes. It was also found that in the course of 12 or 15 months the sides and bottom of ditches usually became solidified from sediment and would sustain very little leakage, and were much easier to maintain.

Water leakage from a ditch was not the only way this precious commodity was lost. Waspish old Adam Shirly, reputed to be the owner of the first water rights on Deer Creek, knew that he was losing water, but was not sure where or how it was happening. Apparently he must of had his suspicions that the problem wasn't seepage because his inspection tours to his ditch were conducted with a rifle over his arm. On a warm July morning in 1865, near Pleasant Flat, three miles southwest of Nevada City, near the road to Grass Valley, he discovered the problem. At this place he came upon a large group of Chinese, by his estimate, at between forty and fifty men, diverting water from his ditch. Without hesitation Shirly opened fire, killing one and wounding another. The matter was taken to the local court where it was quickly determined that the killing was justified. Shirly was released from custody.[44]

As part of the ditch system, feeders were constructed leading off the main channel into storage reservoirs for the purpose of ensuring a constant supply of water and, also, as a back-up in case of an accident to the system. At mines that had an insufficient water supply to maintain a 24-hour operation, or where the stream flow was less than required to maintain maximum operations, distributing reservoirs were also constructed at a convenient location by the mine owner.

All major ditch companies constructed dams to create large reservoirs as the principal backup for their water supply. Most dams in forested areas were built of a crib-work of cross timbers. These were laid upon sills, and placed in hitches cut in the bedrock at a point where a solid foundation could readily be obtained. The crib was then filled with boulders, sand and rock, and in time the material, brought down by the river, would fill in and solidify the dam and make it a permanent structure. For moderate sized reservoirs, generally not exceeding 60 feet in height, earthen dams were frequently constructed designed to have no less than a ten-foot width at the top. The best material for an earthen dam consisted of a combination of gravel, sand and clay properly proportioned giving weight, cohesiveness, stability and imperviousness to the structure. According to mining engineer A. J. Bowie, masonry dams were not justified for hydraulic mining, because of the great expense involved. He did allow, however, that they could be used if the reservoir was designed for other and more permanent uses. It is obvious that Bowie correctly felt hydraulic mining had a limited life expectancy in California, but incorrectly that the stored water would have no further use after its demise. The abundance of reservoirs or lakes in the Sierra today, sired by the demands of the hydraulic mines, testify to the error of his foresight.[45]

The South Yuba Canal Company owned seven large reservoirs: Fordyce, Meadow Lake, Lake Stirling, Cascade Lakes (three), and a dam at the North Fork of the South Yuba. Fordyce was by far the largest of the seven, actually it was the largest artificial lake in the Sierra, with a dam 75 feet high, 645 feet long, and 90 feet wide at its base built on solid granite. It was faced with granite blocks over wooden stringers bolted firmly together which backed up a lake of 750 acres and was constructed at a cost of $150,000. None of the water was drawn out of South Yuba Canal Company's reservoirs until the rivers and streams were completely depleted. It was the policy of all large ditch companies to hold back the water from their lakes until all other sources were exhausted. Then, as in the instance of the South Yuba Canal Company, in a little over two months time, four billion gallons of water would be drawn from all the lakes and reservoirs in their system. For a better perspective of the magnitude of these operations, it should be noted that this amount was calculated to be enough water to furnish each inhabitant of the City of San Francisco (in 1882) with 50 gallons of water per day for a period of 320 days.[46]

One of the first major projects completed by the South Yuba Canal Company was the construction of a ditch from the South Yuba to the head of Deer Creek and to the Alpha and Omega mines. In the process of construction the company blasted through a bluff of solid granite for over a mile, the cliff in places was over 80 feet in height forming a shelf 15 feet wide. A tunnel was also driven through the Deer Creek and Steep Hollow Ridge, 3,100 feet in length, 204 feet from the summit. The largest ditch of the South Yuba Canal Company was constructed from the head dam on the South Fork to the Bear River, transmitting 7,000 inches of water through a canal six feet wide at the top, and five feet deep with a fall of 13 feet to the mile. The entire system of ditches supplied over 200,000,000 gallons of water daily. They ranged in length from two or three miles to 70 or 80 miles. Costs for ditch construction of this kind varied according the obstacles to be overcome and could range from $4,000 per mile to a much as $20,000 per mile.[47]

By 1876, the South Yuba Canal Company owned 300 miles of canals and flumes. The Dutch Flat branch, which was begun in 1864 and completed the following year took its water one and one-half miles below the head of the main channel and ran a distance of 23 miles. The hydraulic mines of Gold Run and Dutch Flat in Placer County, received 70,000,000 gallons daily through this ditch.[48]

The Middle Yuba Canal Company was begun in 1853 by M. F. Hoit of Nevada City, and was completed three years later. It took its water from the Middle Yuba River a short distance above the mouth of Bloody Run. The main ditch carried water to Badger Hill, North San Juan, Sebastopol, Sweetland, Birchfield and French Corral, a total of 40 miles. By 1867 it had a capacity of 1,500 inches and was constructed at a total cost of $400,000, (some sources say $600,000).

Its rival, the Eureka Lake Company, had a main channel 75 miles long with 190 miles of branches for a total of 265 miles of ditches. This system of ditches, including the storage facilities, represented an expenditure of $900,000. Its principal reservoir was Eureka Lake, also known as Canyon Creek Lake, or French Lake which covered an area two miles long and one mile wide. Near the summit of the Sierra was the location of Lake Faucherie and four smaller lakes which comprised its major facilities for water storage.

The most spectacular part of this water system, and a significant engineering achievement, were two magnificent aqueducts—the Magenta, 126 feet high and 1,400 feet in length, and the National, 1,800 feet long with a height of 65 feet. The flume, which conducted water over these giant structures was seven feet wide, but the sides were only 15 inches high. Mr. Benoit Faucherie, the engineer, designed the flume in this manner to keep as low a profile as possible so that the high winds that blew across the gap would have little effect. The North San Juan *Hydraulic Press* has left us this impressive eye witness account: "The entire structure, as one views it from a short distance, looks massive yet graceful and as though it could not be overturned by anything less powerful than a hurricane. It is not built in a straight line across the Gap, but sinuously, following the windings of the ground, which is to increase its strength. The box is not sided up with boards, but with solid stringers."[49] The cost of these imposing structures, which were separated by a small hill, was reported to be about $10,000. Upon their completion the event was celebrated by a magnificent festival, unlike anything ever before witnessed on the San Juan Ridge—even attended by the governor of the state. During the ceremonies, both the flag of the United States and France was displayed, accompanied by a brass band proudly playing the Stars Spangled Banner followed by the *Marseilles*.[50]

The Eureka Lake & Middle Yuba Canal Company, out of mutual need, consolidated their holdings which together were then valued at $2,250,000, including 16 distributing reservoirs. In addition to their extensive water rights, they also owned some of the most valuable gravel mining land in Nevada County, including properties at Columbia Hill, North Bloomfield (adjacent to the North Bloomfield Gravel Mining Co.) and Relief Hill. Also, 120 acres at Snow Point in Eureka Township and 150 acres at Cherokee in Bridgeport Township.[51] By 1883, the water supply of the Eureka Lake & Yuba Canal Company Consolidated was the most extensive in the state with a system of ditches augmented by a series of reservoirs 6,000 feet above sea level. The largest was the Eureka Lake followed by Lake Faucherie located about 600 feet below and two miles distant from the former. The next in size were the two Weaver lakes situated on the tributaries of the Middle Yuba. The dam at Eureka Lake was 250 feet long from bank to bank with an 86-foot base standing a little over 68 feet high and covering 337 acres. By comparison, Lake Faucherie covered 139 acres. The extensive system of ditches delivered 8,800 miner's inches in wet weather and 4,600 during the dry season.[52]

As late as 1861, historian John S. Hittell noted that in El Dorado County hydraulic claims could only be worked during the spring and winter due to the lack of an adequate system of ditches in the region. These same conditions even curtailed work in the drift mines where vast amounts of dirt had to be piled in dumps while the miners awaited the rainy season.[53] To remedy this situation, the California Water Company, with headquarters in San Francisco, brought water to a rather vast area of auriferous gravel deposits located between the Middle and South Forks of the American River. The storage for this water supply came from an extensive system of reservoirs and high dams located in the Sierra at 6,000 feet elevation. It was conducted to the various mines in the region through a series of over 250 miles of ditches, pipes, tunnels and flumes at a cost of $600,000.[54]

But even this system was inadequate to service all of the great gold bearing areas in the county. According to Rossiter W. Raymond, the rich gravel deposits at Kentucky Flat, as late as 1872, were being mined with rather primitive methods utilizing a very limited water supply, varying from 20 to 100 inches. Small hydraulic operations were still being worked with duck hoses with short strings of sluices which were operated without even the benefit of quicksilver. At this time the mines were producing only about $6.00 per day per man.[55]

In other parts of the county, it appears, water was not such a limiting factor. On the Tertiary channel of the South Fork of the American River between Texas Hill and Coon Hollow, during the decade of 1861 to 1871, $10,000,000 was produced by hydraulic and drift mining operations.[56]

It is also worth noting that in the productive hard rock mines at Georgia and Spanish Dry Diggings in El Dorado County, in the vicinity of Georgetown, the seams of gold in the slate formations were sometimes worked by the hydraulic process.[57]

The El Dorado Water & Deep Gravel Mining Company's ditch was not finished until 1876. During its construction, as many as 1,900 men were kept busy speeding its completion. The capacity of the canal was 12,000 inches for 24 hours, however, the company only used 3,000 inches to operate its own mines and, therefore, was able to sell the remaining 9,000 inches to other mines along the course of the channel. The canal, after an expenditure of $600,000, was large enough to float lumber nearly to Placerville.[58]

With the completion of the Park Canal & Mining Company's 290 miles of ditches, carrying 2,200 miner's inches of water, El Dorado County could finally lay claim to the finest water system in the state. Constructed principally from the consolidation of earlier ditch companies, this gigantic operation, formerly named Diamond Ridge, was the result of the enormous expenditure of $2,000,000. The geography of the county was largely responsible for the effectiveness of its ditches. The Middle and South forks of the American River together with the various forks of the Cosumnes and a myriad of high mountain lakes, formed an unsurpassed natural water system. By the 1880s the county was essentially serviced by three major ditch companies. Its topography was such that water could be carried by ditches to nearly every point within its borders from three great ridges or divides: the system of the California Water Company on the North Divide; the El Dorado & Deep Gravel Company's system on the Middle; and the Park Canal & Mining Company's system on the South Divide.[59]

The Dutch Flat-Gold Run mining district in Placer County was located almost on top of one of the richest gravel channels in California—the so-called Blue Lead. With the advent of hydraulic mining it became one of the great mining centers in California. By 1867 there were 45 hydraulic mines operating within a radius of one and one-half miles. As noted earlier, this area was serviced by a main branch ditch of the South Yuba Canal Company, but even that tremendous supply was not adequate to fill the needs of so many mines.

The Dutch Flat Water Company, which owned both The Placer County Canal and the Canyon Creek-Little Bear River Ditch, each about 30 miles long, served the two towns with a respectable capacity of 5,500 miner's inches. Although the water supply for these ditches was fed by an impressive array of twelve reservoirs, water was generally only available from the middle of December until the middle of August. Water rates in 1873 ranged from nine cents per inch for twelve-hour water to 15 cents for 24-hour water. The district was also served by the Auburn & Bear River Canal Co. which owned main ditches and branches totaling 290 miles in length, constructed at a cost of $670,000.[60]

When Rossiter W. Raymond visited Dutch Flat and Gold Hill in 1872, he was greatly impressed by the efficiency of their mining operations, noting the district had achieved considerable success in its efforts to entice outside capitalists to invest in a number of local hydraulic mines. The Gold Hill Hydraulic Mining Company had been purchased in the late '60s by an English company with headquarters in London. Although owning no water rights, it operated with remarkable success simply by purchasing all its water on the open market. The company's improvements included: 1,400 feet of flume, four feet wide and two feet deep; 1,500 feet of 16-inch pipe; 1,500 feet of eleven-inch pipe. Two monitors were worked with a pressure of 240 feet bringing an average yield of $6,000 to $8,000 for a 30-day run.[61]

Nearby, the Dutch Flat Gravel Mining Company in 1871, installed one of the most impressive supply lines in the California mines. It stretched down a hill, in a straight line for 3,500 feet with a huge five-foot diameter funnel at the receiving end tapering down to 22 inches at the tail. It was designed to withstand a pressure of 465 feet, delivering an amazing 3,000 inches of water. In 1875, the owners contracted for the construction of a bedrock tunnel at a cost of $30 per lineal foot. When completed, it was six feet wide and eight feet tall with the head sunk 68 feet below bedrock and 148 feet below the top of the gravel, making possible an 80-foot bank. A perpendicular shaft was sunk through the gravel and part way into the rock where it connected with an 80-degree incline shaft leading from the head of the tunnel. The grade of the bed was ten inches to every twelve feet and led to the Bear River. A four-foot sluice was laid in the tunnel with an additional length of 384 feet laid from the mouth along a steep hillside where a bed was excavated out of solid rock.[62]

The Forest Hill Divide, lying between the North and Middle Forks of the American River in Placer County, could be described as a plateau or tableland of eleven square miles which at one time contained 16 important mining camps: Forest Hill, Iowa Hill, Menona Flat, Damascus, Sucker Flat, Wisconsin Hill, Michigan Bluff, Todd's Valley, Yankee Jim's, Last Chance and five or six others of lesser note. Although some of these mines, such as those at Yankee Jim's and Iowa Hill, were among the very first to adopt the hydraulic method, because of a chronic shortage of water, most mining in the region was carried on through drifting.[63]

The first major company to bring water to the district was the Iowa Hill Canal Company organized in 1872, however, for some years previous to this date, the region had placed its hopes in a grandiose scheme designed to bring water from Lake Tahoe. This plan was the so-called Von Schmidt project which was conceived to supply water to a good portion of Placer County. Colonel Alexis Waldemar Von Schmidt, a brilliant but somewhat starry-eyed engineer, in 1865, founded the Lake Tahoe & San Francisco Water Company, purchasing a half section of land surrounding the Truckee River outlet on the north shore of Lake Tahoe for $3.00 per acre. He likewise obtained water rights on the Truckee River amounting to 500 cubic second feet of water. By the fall of 1870 Von Schmidt had completed a crib dam strategically located 150 yards below the lake's outlet. His overall plan was to run a combination railroad tunnel (for the Central Pacific) and aqueduct through the mountain near the entrance to Coldstream Canyon, eliminating 20 miles of railroad snowsheds. Von Schmidt also planned to furnish San Francisco with unlimited supplies of water with branch ditches designed to serve a vast mining area in Placer County, including Yankee Jim's, Forest Hill and Michigan Bluff. However, Von Schmidt's impressive scheme was doomed to failure. The City of San Francisco would not seriously consider the plan nor was the Central Pacific interested— in short the plan was found to be both impractical and not cost effective.

By 1879, the Iowa Hill Canal Company had completed 27 miles of main channel, with an excellent capacity of 8,000 inches. The ditch measured eight feet at the bottom, eleven feet at the top with a depth of four feet on a grade of ten feet to the mile. Upon its completion it was one of the leading canals in the state with 45 miles of branch ditches and the capacity ultimately increased to 10,000 inches of water. [64]

In a study commissioned by the Iowa Hill Canal Company, consulting engineer, R. H. Stretch, determined that although the charge for water to the miners in this area varied from 12 to 25 cents an inch for 24 hours, the cost to the ditch owners was only three to five cents per inch. At this figure, Stretch concluded the ditch owners could work gravel profitably, on their own account, which would prove ruinous to their customers. He recommended that the company operate their own hydraulic mines in areas that would otherwise be considered marginal ground to anyone else.[65]

Gold was found in all parts of Sierra County, a land marked by deep canyons, steep peaks and fast-flowing rivers wedged between Nevada County to the south and Plumas to the north. Although vein gold was found in abundance, its rich Tertiary gravel deposits made it the leading county in California for the bullion taken from its many drift mines. However, as more water systems were developed, hydraulic mining became an increasingly important method of working the deep diggings. By 1872 there were at least 50 mining ditches in the county, with an aggregate length of 220 miles, costing in excess of $750,000. However, even then, the water supply was only adequate to maintain the mines for a period of four to eight months of the year.[66]

The greatest area of concentration for hydraulic mining was located in the northern part of the county where, until 1866 (when the boundary was changed), it was bordered to the north by the South Fork of the Feather River and bisected by Slate Creek, an important tributary to the North Yuba River. To "old timers" in Sierra County, even today, this rather remote area is known as "Over North." According to Sierra County historian James Sinnott, this region had once contained more settlements of significant size than any other part of the county. In the spring of 1850, rich placers were located along the banks of Slate Creek, bringing the inevitable rush of miners to the area. In rapid succession dozens of mining camps mushroomed along this waterway bearing such colorful names as Whisky Diggings, Howland Flats, Gibsonville, Poker Flat, St. Louis, Port Wine, Scales, Poverty Hill and Rabbit Creek to name but a few.[67]

The main Tertiary gravel deposit, known as the La Porte Channel, extended southwest from Gibsonville into the La Porte mining district (Rabbit Creek). It then continued in the same direction until it was joined by a branch leading from the St. Louis-Table Rock region. Howland Flats, for a number of years, shipped over $600,000 annually from both its drift and hydraulic mines. The Fair Play hydraulic mine was a big producer and had been worked since 1859. By 1882, its owners, the Boyce Brothers, were working a bank 400 feet in depth. The main branch of the Tertiary gravel deposit continued into the Poverty Hill-Brandy City district where by 1870 the mines in the vicinity had produced over two million dollars in gold from easily worked ground. It then became necessary to construct costly bedrock tunnels to work the deeper gravels.[68]

One of the earliest ditches in the area was the Sears Union Water Company ditch which was begun as early as 1853. The company was organized to convey the waters of Slate Creek and tributary streams to Sears Diggings, later renamed, St. Louis. Beginning with a capital stock subscription of $75,000, in time it became the leading ditch company in northern Sierra County. Including its branches, it was

18 miles long with two miles of fluming. It would eventually service the mining camps of Howland Flats, Pine Grove, Potosi and Cedar Grove. The charge for water during most of the '60s was 30 cents for a 24-hour inch delivered for hydraulicking and 50 cents to drifting companies.[69]

The ditches of Sierra County were all relatively short and obtained their water supply from creeks and springs, with very few substantial reservoirs. Many of the hydraulic mines, as well as drift mines, received their water from their own individual ditches. At Gibsonville, the Union Water Company operated two ditches that obtained their water from Slate Creek, while two others, owned by the same company, had their source from springs in nearby ravines.

Brandy City, which was the principal hydraulic mining camp in the county in 1868, was then using about 3,000 miner's inches supplied mainly from a number of small ditches, the three largest being: the McGowell Ditch which obtained its water from Fiddle Creek, east of town, later known as the Arnott Ditch, and the Hoosier Ditch, later to become the Brandy City Ditch. It was eleven miles long and took its water from Canyon Creek. Finally, the smallest, the Grizzly Hill Ditch, only two miles long with Cherokee Creek as its source.[70]

Although Downieville had experienced considerable hydraulicking of the easily worked placers in the early going, by 1882 only two mines remained in operation

Sierra County, as we have learned, did not have the great water systems to match those of its neighbors to the south, but even so, the production of the gravel mines of "Over North," alone, from 1855 to 1883, was in excess of $60,000,000.[71]

Farther to the north in Plumas County the big ditch of the North Fork Company, built particularly to work the mines near the North Fork of the Feather River, about six miles south of Big Meadows (present Lake Almanor), was not completed until 1875. The ditch was twenty-five and one-half miles long including eight miles of pipe and a number of tunnels. The Plumas Water & Mining Company owned the most valuable gravel locations and water privileges in the county, however, like its neighbor, Sierra County, the greater part of the deep gravel mining was confined to tunneling while most of the hydraulic mines were limited to the La Porte and Poorman's Creek section of the county. At La Porte the Martindale Ditch later known as the Geeslin Ditch was completed in 1855 and furnished about 1,200 miner's inches. When first completed the company charged $1.00 per inch which was later reduced to 50 cents when the competing Rabbit Creek Flume was completed in 1857. By 1860, in addition to the Martindale Ditch, there were three other major ditches in the area: the Yankee Hill Ditch, the Feather River Ditch, and the John C. Fall Ditch. The largest and richest claim at La Porte was owned by Conly & Gowell which had been developed at considerable expense. This mine had an abundance of water using about 800 inches through an eight-inch monitor. In four months time its owners had washed an impressive sum of 1,380,807 tons of gravel.[72]

Hydraulic mining was introduced into Yuba County at Sucker Flat located near the main Yuba River about one mile above Timbuctoo. The Sucker Flat Channel extended from Timbuctoo to Smartsville and was heavily mined by the hydraulic method. Over the years, many early mining camps along the Yuba River such as Rose's Bar, Saw Mill Bar, Lander's Bar, Kennebec Bar, Cape Horn and Saw Hill were buried by mining debris, some by as much as 200 feet. The Tri-Union Ditch brought water 60 miles from Deer Creek to Sucker Flat, Timbuctoo and Ousley's Bar. The Long Bar Ditch brought water from Dry Creek to Long Bar, a distance of five miles. The Excelsior Ditch, after an expenditure of $250,000 delivered water from the Middle Yuba and Deer Creek, a distance of 32 miles to Mooney Flat, Timbuctoo and Eureka Flat. The leading mines were concentrated in and around Smartsville and included Pactolus, Rose's Bar, the Blue Gravel, Smartsville Hydraulic Mining Companies, the Blue Point, Golden Gate and Young America.[73]

The Blue Gravel Company at Smartsville, for a number of years, was the richest placer mine in the state. Work on this claim began in 1853 with the construction of a drainage tunnel, but as we learned earlier, was not completed until March 1864. During that time $315,489 in gold was washed out, but this amount was exceeded in expenses by $7,543. However, during the next 43 months of operation, the stockholders divided $564,500 in dividends with an additional $61,043 expended in new improvements. During this period 1,600,000 cubic yards of material was washed through the sluices. The cost of water was $567,261 at the rate of 15 cents per miner's inch. Clean-ups were generally made at intervals of two to three months. The sprawling claim covered an area in excess of 100 acres. The bedrock tunnel was 1,700 feet long and was excavated at a total cost of $80,000. In 1863, as the mine approached a greater depth, another tunnel was begun which was completed three years later after an additional outlay of $75,000. The mine operated on 500 miner's inches of water and required 125,000 pounds of powder per year. According to the mine's owners, the incredible amount of three tons of quicksilver was required to hold and save the gold as it passed through four miles of wooden, block-paved sluices.[74]

The oldest, continuously operating hydraulic mine in California—the Depot Hill Hydraulic mine, was located in Yuba County five miles north of Camptonville. It was opened in 1855 and continued every season until finally ceasing operations in 1942. For most of that time, until its closing, it was the property of the Joubert family, handed down for three generations. Before 1927 it had produced over one and one-quarter million dollars with a yield of 20 cents per cubic yard. It was situated on the La Porte-Scales-Brandy City Channel, getting its water through seven miles of ditches from a branch of Indian Creek. The water supply was capable of producing between 600 and 700 miner's inches at 160 feet of pressure. At the height of the season, the Jouberts would operate one five-inch nozzle and either one or two additional four-inch monitors, working a bank 80 feet high and 50 feet wide. After the Sawyer Decision, the debris was held at Bullard's Bar where storage charges ran two cents per cubic yard of material which, was processed through 5,100 feet of sluices. Before 1903, water was available for about 4,300 hours of work, but by 1932 the supply had been reduced to 1,200 hours. When the mine was in full operation the crew consisted of five to six men with Joubert doing the piping. The smaller nozzle was used for cutting and the larger for sweeping.[75]

Water was first brought to the placer mines of Butte County in 1852 by a group of investors bearing the rather cumbersome, but descriptive name of The West Branch of the Feather River Company. It was organized by issuing shares of capital stock in the amount of $200,000 for the purpose of bringing water to the rich placers of Long's Bar. Later that same year the Wyandotte & Feather River Water Company was incorporated to construct a ditch to Forbestown and the hydraulic mines of Wyandotte, Honcut, Ophir and Bagdad. In 1854 these water rights were sold to the South Feather River Company, resulting in the construction of a ditch to Wyandotte the following year. Eventually it would gain the distinction of becoming the oldest ditch still in existence during the height of the hydraulic mining activity. It extended for a length of 30 miles and held prior rights to any other ditch company on the South Feather River. The following winter (1855-56) the Feather River & Ophir Water Company constructed a ditch to Oroville, making that emerging town the leading city in Butte County. In 1858 the Walker & Wilson Ditch was completed to Thompson Flat. Drawing its water from Little Butte Creek, this impressive canal stretched for a distance of 36 miles. By 1861 Butte County could boast eight major mining ditches with an aggregate length of 167 miles. But the most significant ditch project in Butte County was that of the Spring Valley Water Company, destined to become one of the great engineering achievements in the history of California mining. This magnificent accomplishment and the Cherokee Mine it served will be dealt with, in some depth, later in this study.[76]

By 1882, in addition to the Cherokee Mine, there were approximately seven significant hydraulic mines operating in Butte County: The Miocene Mining Company, described earlier in this chapter — with flumes clinging to the side of a shear cliff, served by 34 miles of ditch and over 5,000 feet of pipe carrying 3,000 inches of water; the vast Powers Mine, owned by the Oroville Mining & Irrigating Company, comprising over 1,500 acres of mining ground and 30 miles of pipe; the Hewitt Mine, also owned by the Oroville Mining Company with 22 miles of ditch and 3,000 inches of water; the Hendricks Mine, a few miles from Cherokee and situated at a bend in Table Mountain with an investment of $318,000 in flumes, pipes, ditches and water rights; the South Feather Water & Union Mining Company, located at Forbestown and served by the 30-mile Forbestown Ditch; the Red Hill Hydraulic Company, near Magalia, incorporated in 1872, with four miles of flumes and operating two Chiefs; finally, the latest, the Mineral Slide Mine opened in 1882 on Big Butte Creek.[77]

By 1881 there were a total of 40 mining and 25 irrigation ditches in Butte County, with an aggregate of 701 miles of main channels and branches. At that time it was estimated that 10,000 miner's inches of water were used daily.[78]

One of the earliest ditches to be constructed in Amador County was completed in 1851, it took its waters from Jackson Creek and extended for a distance of seven miles. By 1858, after several changes in ownership, it became known as the Kilham Ditch and, in that year, it reduced the price of water from 75 cents an inch to 40 cents. Although it was profitable in the early going, later, after being purchased by the Butte Ditch Company, it ran into serious financial difficulties eventually forcing a sacrifice sale. Other Amador County ditches also proved to be less than successful. The so-called Ham Ditch, constructed by J. C. Ham, designed to serve most of the western and middle portions of Amador County, required eleven and one-half miles of costly flume and other unanticipated expenses resulting in a disastrous failure to its investors. It was finally acquired, after a considerable loss, by Pioche, Bayurque & Company of San Francisco. The Willow Springs Ditch proved no exception and was also a financial flop. Begun in 1851 and flawed by poor engineering it was eventually sold to D. O. Mills & Company of Sacramento who, after a complete redesign, operated it profitably as a lumber flume.[79]

Although, as we have seen, the history of ditch building in Amador County, for the most part, was a litany of one failure after another, the construction of the Amador Canal, completed in 1875, gave rise to the optimistic notion that heavy gravel mining would become a notable factor in future years. Built on the remains of the bankrupt Sutter Canal & Mining Company's ditches and the water-rights of the defunct Butte Ditch Company, the new canal took its waters from the main fork of the Mokelumne River. From there it ran for a distance of 45 miles, carrying 5,000 inches of water in an impressive ditch measuring six and one-half feet wide on the bottom, nine feet on top and three feet deep, serving a considerable portion of Amador County. While the benefits of cheap water power led to the reopening of a number of moribund quartz mines and added to the prosperity of others which earlier had relied on the more costly expedient of steam pressure, the opening of extensive hydraulic mining never became a reality.

Toward the end of 1875, the Amador Canal Company opened one new hydraulic mine about four miles from Jackson, known as Martell's Claim which, for a while, became the largest in the county, working two monitors. For years, there was small-scale, desultory hydraulicking from Plymouth up to Volcano notably on the headwaters of Rancheria and Dry Creek and also at China Gulch. However, even with ample, cheap water from the Amador Canal, the ground was never rich enough to properly develop an important hydraulic mining operation, with the significant exception of the Elephant Hydraulic Mine located above and north of Volcano. It is interesting to note that this mine was one of the last to shut down in California and was operating on a small-scale lease arrangement well into the 1930s. Modest hydraulic operations were also carried on for a time on the channel east of Pine Grove.[80] As mentioned earlier, notable tertiary deposits were limited in most of the Southern Mines.

Calaveras County could possibly be named the one exception. By 1857 the county could boast 17 ditch systems, 325 miles in length. As we learned in an earlier section, rather extensive Tertiary channels ran through the county, however, for the most part they were mined by drifting. On the Fort Mountain channel system, for example, the two branches of the ancient Calaveras River were worked mainly by tunneling with some small hydraulic operations. At the Baldwin hydraulic mine near Calaveritas, 80 acres were worked by hydraulicking on a gravel bank 75 feet in height, with a width of about 600 feet and length of 2,640 feet. The El Dorado channel near

Mountain Ranch was also hydraulicked at the Rose Hill, Humboldt and Motto pits. Although from Sheep Ranch to Calaveritas, erosion had removed much of the channel, at Cave City and Old Gulch, profitable hydraulic operations were established. By the 1880s the Mokelumne Ditch & Water Company operated a hydraulic mine on Chili Gulch utilizing 200 acres of property. The Calaveras Hydraulic Mining & Water Company worked extensive beds of gravel six and one-half miles north of Milton. The ground was 400 acres in extent washed with 1,500 miner's inches of water from two giants equipped with six-inch nozzles. This property, which generally employed up to 25 men, operated under various names from early 1871. After the Sawyer Decision, three impounding dams were installed, allowing limited operations.[81]

Among the larger ditch companies in Calaveras County, in addition to the Calaveras Mining & Water Company, the Union Water Company with an investment of over $200,000 and the Campo Seco Company with operations costing in excess of $500,000, should be included. In the Comanche-Milton area on the Ione formation, numerous hydraulic mines operated for a number of years, particularly in the lower section. In all, over 50 miles of ancient river beds were located in the county.[82] By the mid-'70s, renewed activity in placer mining prompted the *Mining and Scientific Press* to report: "Calaveras county now rivals the more northern counties in hydraulic mining. New locations are constantly being made, and various tunnel and hydraulic operations are vigorously prosecuted. The mines are of a permanent character, and the result of the present season's work will be a good one. Some of the mines now worked were a few years since considered as exhausted diggings."[83]

In nearby Stanislaus County, the La Grange Ditch & Hydraulic Mining Company was organized in 1871, taking water from the Tuolumne River at Indian Bar. The length of its ditch totaled 25 miles and was constructed at a cost of $450,000. Many hydraulic mines operated in this region between the '70s and '80s. Although the La Grange Mine managed to stay in operation with moderate success until the Sawyer Decision in 1884, it should not be confused with the great La Grange hydraulic mine located in Trinity County.[84]

WATER, DITCHES, AND DAMS

Hydraulic mining was hardly confined to the so-called Gold Rush country, but was found in other parts of California as far south as the San Gabriel Mountains in Los Angeles County, where the Texas Point Mine, in the Lytle Creek District, produced $50,000 in a single year. Or as far east as Dogtown and Monoville, in distant Mono County, where water was brought from a great elevation through a 20-mile ditch from Virginia Creek. There in the early '70s the 1,700 by 200-foot Sinnamon Cut produced $80,000 from gravel washed through its sluices using the hydraulic process. Scores of small hydraulic operations were also carried on successfully in Humboldt, Siskiyou and Del Norte counties, particularly in the Klamath Mountain area bounded on the north by the Klamath River and south by the Trinity River and the forks of the Sacramento. Nor should Shasta County be ignored, there small hydraulic operations were noted as early as 1856 and mines such as the Piety Hill Blue Gravel, located at Igo southwest of Redding, was described by Raymond in the early '70s. Supplied by 800 inches of water through a 23 mile-long ditch, the mine was washing ground with values of 37 cents per cubic yard. But the major thrust of hydraulic mining was in the the seven gold producing counties, as noted earlier, bounded by the North Fork of the Feather River and the South Fork of the American River. Trinity County, a major exception, will be treated separately in a later chapter.[85]

To put the distribution of hydraulic mining operations, throughout the gold regions of California, in proper perspective, it might be helpful to note that of the 50 leading hydraulic mines, only one was located south of the Cosumnes River—the Elephant in the Volcano District of Amador County.[86]

Water consumption from the ever-increasing size of the nozzles coming into use, some with a capacity of over a million gallons an hour, was simply overwhelming. It became impossible to keep up with the demand. In the spring of 1871, even the Nevada City mining districts, the site of the first miner's ditch and headquarters for the South Yuba Canal Company, was particularly hard hit by a scarcity of water. On March 11, the *Daily Transcript* reported that some of the hydraulic mining companies, including the Walkenshaw and the Eagle had shut down their nozzles and were resorting to drifting as a viable alternative. By this method, it was learned, these mines were paying about $30 to the hand.[87]

The following week the *Transcript's* readers were informed,

> The demand for water to be used for mining purposes is greater in Nevada Township this season than for years before. The South Yuba Canal Company has demand for 1,800 inches per day, or more than all the ditches supplying this locality can convey. The entire sales of water by the South Yuba Canal Company, last week, was 31,000 inches and the largest week's sales ever made by the company was only 52,000 inches. This could probably be sold now if it could be supplied. . . . The ditches which supply this vicinity carry 500 inches less than there is a demand for.[88]

A few days later the water shortage had caused the mines to make even further accommodations, ". . . in many of the mining camps the miners have so arranged their operations that they are enabled to alternate in the use of water—some of the companies working in the day time and others working at night." The column added: "The demand for water is much larger than ever before and is probably owing to the fact that larger heads of water are run and hydraulic mines are worked on a more extensive scale."[89]

In response to the growing demand for water, new ditch companies were making an appearance and older ones were expanding their operations. In addition, large hydraulic mining companies, such as the North Bloomfield Gravel Mining Company and the Milton Mining & Water Company were constructing their own ditches with the expectation of selling surplus water. On June 17, 1874 the *Transcript* printed a letter from their correspondent at Moore's Flat on the San Juan Ridge: "The Milton Company have started to work on their ditch and will rush it through as fast as possible. They have four hundred Chinamen employed and are employing all of the white men they can get. . . . A genuine 'blow out' will be had here the day they break ground on this side of the ridge and Champaign will flow like water."[90]

Two months later the same paper gave an update on the progress of the Milton Ditch, revealing there were then 1,500 men employed on the job, 1,300 being Chinese (always a touchy subject with mining town newspapers). The ditch was being worked in four sections and progress had been made to a point within four miles of Moore's Flat. The item continued: "The ditch, when completed, will carry three thousand inches of water. It will be 30 miles in length, [it would actually be 80 miles long ending at French Corral] and will be one of the finest built ditches in the state." The Milton water system, when completed extended from Rudyard Reservoir to French Corral and represented a total expenditure of $391,579. With respect to the North Bloomfield Gravel Mining Company, which will be discussed more fully in a later chapter, the length of its ditches, including reservoirs, totaled 157 miles with a construction cost of $708,000.[91]

The source of the water supply of the North Bloomfield Gravel Mining Company, the Milton Mining & Water Company and the Eureka Lake & Yuba Canal Company was at the headwaters of Big Canyon Creek and the Middle Yuba River in Nevada and Sierra counties. The catchment sections alone embraced an area of 68.6 square miles, containing eleven principal reservoirs, totaling 11,600 acres, with a capacity of over 2,195,000,000 cubic feet. The great gold-bearing area, defined by the North Fork of the Feather River and stretching southward to the environs of the South Fork of the American River, which encompassed the richest hydraulic mining counties in the state, contained the enormous aggregate storage capacity of 50 billion gallons.[92]

Water was the greatest single expense for the hydraulic mine owners and could spell the difference between profit or loss. In 1874, for example, when the price of water at North San Juan, with the completion of the Milton Ditch, was reduced from 16-1/2 cents an inch to eight cents, the American Mine at Sweetland was able to save $150,000 in a single year from the reduction alone.[93] Because of the critical relationship between the cost of water and profit or loss, as might be expected, there was a constant war between buyer and seller of water as to the true quantity supplied or the integrity of the miner's inch. It was generally known that water in motion towards the aperture would run more than "dead" water. Also pressure was difficult to keep steady. The least variance in the shape of the aperture had a rather profound effect. For example, if beveled a little favorably, it was possible to increase or decrease the volume of water by as much as one half.[94]

In an earlier chapter it was noted that the first pipe to be used in a hydraulic mining operation dated back to 1853, and consisted of 100 feet of stove pipe. From this early beginning, the use of pipe spread rapidly, mainly for the purpose of bringing water to the individual mine sites. However, it should be noted that the use of hose, as a distribution conduit to the nozzle, was in wide use well into the '60s. At that time hose was being advertised as withstanding pressures up to 200 inches and was available in sizes ranging from four and one-half inches to eight inches.[95] Nevertheless, in 1856, the Union Iron Works of San Francisco began the manufacture of wrought iron pipe to be used in the hydraulic mines and as early as 1857, a large sheet iron pipe, 40 inches in diameter, was laid as a water conduit at Timbuctoo in Yuba County.[96] Wrought iron pipe was used extensively because of its weight, cheapness of construction and tensile strength. For convenience and economy, large hydraulic mining companies began forming their own pipe in company-owned shops. They purchased the sheet iron and then fabricated it into pipe by a process of rolling and riveting. The material varied in thickness depending on its use, but generally ranged from one-sixteenth to three-eights of an inch.

In the earliest days, it was joined together much like stove pipe. However, for extremely high pressures, an iron collar or a lead joint was used. Pipe lengths usually ran from 20 to 30 feet, with diameters ranging, generally, in 11-15-22-30 and 40-inch sizes. Sheet iron was purchased from the supplier in sections of 30 to 36 inches and then riveted into the desired lengths. Usually a single row of rivets were run in a set from one to one and one-quarter inches apart on the horizontal seams and from two to three inches apart on the vertical seams. For unusually high pressures, double rows of rivets were installed.[97]

For protection against rusting and corrosion, sheet metal pipe was painted with asphalt inside and out, giving it a life-expectancy of

25 years. For maximum efficiency, large sections of iron pipe were dipped into a bath of heated asphaltum. It was found that a comparatively light sheet metal pipe, in sizes of moderate diameter, when properly proportioned to diameter and pressure, was both cheaper and more satisfactory than other pipe. Upon installation, care was taken to lay the pipe upon a solid foundation along its entire length. Where possible, it was laid in a trench and covered with earth for protection from expansion and contraction or injury from slides, boulders or falling trees. It became good practice to lay the pipe from the discharge end and from there work uphill. From sad experience, it was learned that when starting at both ends, no matter how accurate the measurement, when making the final connection, gaps could be caused simply by changes in the temperature.[98]

The dilemma of expansion and contraction was the greatest problem in the use of pipe for hydraulic mining operations. It was learned that all pipe expanded or contracted to some degree as it was being laid, often pulling slip joints apart. To remedy this unhappy situation, a special coupling was invented and patented as the "Loveridge Expansion Joint."[99]

Although large earthenware water conduits had been used by the ancient Minoan civilization as early as 1300 B.C., the use of wrought iron for this purpose was very limited until adopted in California. Thus hydraulic mining set in motion a whole new industry. The Union Iron Works of San Francisco was founded in 1850 and built its first plant at Ist and Mission Street, later becoming the state's greatest supplier of wrought iron pipe and claiming the distinction, for a time, of owning the only steam riveting machine on the west coast. The increasing need for mining equipment for both hydraulic and hard rock mines stimulated the formation of other foundries and iron works in San Francisco. Most were established in the Happy Valley section of the city, and included the Eagle Iron Works, Alta Foundry, Vulcan Foundry, Pacific Foundry, Sutter Iron Works, and the Risdon Iron Works.[100]

During the construction of the system of ditches that supplied the mines in the Sierra foothills and elsewhere in the state, it was often necessary to cross valleys, gulches and canyons too deep to build a trestle. Solving this problem became the greatest challenge in the use of iron pipe, and marked another of the engineering achievements sired by the hydraulic mines of California. This accomplishment brought into use the inverted siphon, which was based on the simple principle that when water enters the pipe it must have a higher head than when leaving it. When making such an installation, extremely high pressures could be generated, requiring air valves to be affixed at intervals to allow air to escape when the pipe was being filled, but more importantly, to prevent the collapse of the pipe in the event of a break.

In the construction of an inverted siphon, the pipe could be supported by suspension cables, or as in the case of a design engineered by Joseph McGillivray, purported to be first to use the technique, a suspension bridge was actually constructed across the Trinity River, in 1869, for the purpose of supporting a 15-inch wrought iron pipe. Water was brought into the pipe from a ditch located at an elevation of 240 feet above the bridge. The pipe was 1,980 feet in length with the outlet 133 feet below the inlet. The McGillivray Ditch took water from Canyon Creek four miles above the mouth of the ditch which was fed by a dam below Reservoir Hill. The suspension structure also served the shrewd McGillivray as a toll bridge across the Trinity River where it became part of the road system from Junction City to the North Fork.[101]

By 1883, there were no less than 6,000 miles of major ditches and at least 2,000 miles of minor branches and pipelines, conducting water from reservoirs with reserves of over seven billion cubic feet and valued in excess of $100,000,000. Much of this system is still in use today, a legacy of the hydraulic mining era, serving the communities of the Sierra foothills with domestic water, irrigating the great agricultural regions of the Central Valley and generating hydroelectric power for the Pacific Gas & Electric Company throughout northern California.[102]

CHAPTER 4

Invention and Maturity

By the year 1868, hydraulic mining had arrived near the brink of its greatest development, and before the end of 1872, it had reached maturity. Remarking on the state of the industry in that year, Rossiter W. Raymond observed that hydraulic mining had witnessed a revolution in new and improved appliances "by which this branch of mining has become a regular and legitimate business."[1] The long recession in the mining industry was all but over. Drift mining was becoming prosperous as miners began to learn the proper levels to run their exploratory tunnels. Also, lode mines, which had previously been shut down, were now reopening where new and successful operations had been inaugurated. Finally, hydraulic mining, with large systems of dams, ditches and reservoirs, was becoming quite profitable.[2]

But even with the obvious recovery in the California gold mines, the days of '49 and the early '50s were gone forever. The shallow diggings in the rivers and streams and along their banks and bars had been exhausted for over twelve years. Plenty of gold was still in the ground, but by then and for years to come, it would be worked by massive capital investment rather than intensive labor. Thus, many miners were forced to turn to other occupations. The population of the twelve principal mining counties was reduced from 140,000 in 1860 to 113,000 in the following ten years. However, of even more significance, of those that remained, only half as many were engaged in mining as in 1860.[3] As Charles G. Yale commented in a report to the American Institute of Mining Engineers, ". . . the character of the mining and the character of the mining population changed. It was no longer possible for a nomadic miner with pick and pan, to gather a fortune in a few days. . . . It became necessary to employ capital, as well as labor. . . . Companies took the place of the individual miner, built ditches to bring in water to the gravel claims, and mills to crush the ore. . . . The miners gradually stopped working for themselves, and were employed by the companies for daily wages."[4]

Personal income for the placer miner also reflected this change in status. A day's wages in 1848 were pegged at 20 dollars, and in '49—16 dollars. While there was very little hired labor during this period, these figures represented the average daily gold yield per miner. As the easily worked placers became more scarce, the per-capita yield, likewise, progressively declined each year. In 1850 ten dollars was considered the going wage, dropping to eight dollars the following year, and to six dollars in 1852. Before the end of 1853, the year of the introduction of hydraulic mining, they had declined to five dollars. By 1856 they stabilized at three dollars per day and remained in this area until 1879.[5] As the development work in the hydraulic mines decreased in the latter part of the '70s and the system of ditches and dams were completed, there was a growing surplus of labor near the end of the decade.

In 1879, when the Derbec drift mine at North Bloomfield cut the miner's wages from three dollars for a ten hour day, to two and a half dollars for an eight hour day, the working force actually walked off the job, closing the mine. But with no union and little organization the strike was soon broken and the mine reopened with the new wage scale firmly in place. A miner's meeting was called in Nevada City and an attempt was made to form a union, but after several fruitless meetings the endeavor failed and other gravel mines, both drift and hydraulic, began instituting the new scale of $2.50 per day *for a ten hour day.*

But of even greater concern to the white labor force was the competition of the Chinese miner. For years the growing enmity toward mine owners who hired "Chinamen" had given rise to increasing cases of vandalism on company-owned property. These costly acts of destruction and the swelling bitterness of popular sentiment fanned by inflammatory mining town journalism, finally led some mine owners to yield to public pressure. On May 18, 1879, one local newspaper chortled: "Major Downie, Superintendent of the Centennial mine, has discharged all the Chinamen until now employed by him and will hereafter have only white men in his camp. This is a move in the right direction. If the remaining Superintendents in the county will follow the Major's example, the problem of Chinese immigration will be effectually solved."[6]

But we have gotten ahead of our story.

By the late '60s there was an expansive spirit of optimism in the gold regions of California created, in no small part, by the progressive leadership of Nevada, the chief gold producing county of California. It was there that California quartz-mining and milling had its crude beginnings and later development. It was Nevada County, also, that saw the first use of the sluice, and the first ground sluice, and where hydraulic mining and drift mining had their origin and chief development. It was there, too, that the first quartz-mining district in America was organized, and the first district laws regulating quartz mining adopted.[7] The positive influence of this richly endowed and unique county was given effusive expression by a traveling correspondent for the San Francisco *Alta* in an article written in 1867.

> Crossing Bear River I came into Nevada, the leading mining county—far in advance of all the others for the variety of its mines and the amount of gold production. It has more quartz mills, more miles of ditch, more sulphuret works, more miners, larger towns, and more capital invested in mining than any other county. It has always been distinguished by the superior intelligence and enterprise of its population, and most of the improvements made in mining in the State have their origins in Nevada.[8]

The 1870s were definitely the halcyon years for the hydraulic mining industry. The first half of the decade witnessed unprecedented technological advances, leading to continuing development, growth and profits. It would not be until the latter part of the '70s that the debris controversy would begin to cast a dark shadow over the industry, raising forebodings for its future. Although Thomas Starr King noted as early as 1860, that the rivers were already perceptively affected by sediment from hydraulic mining, the miners, as late as 1874, were not taking the farmer's protests seriously. In a reply to a demand by the Wheatland *Enterprise* that hydraulic mining be stopped, the Nevada *Daily Transcript* glibly replied: "We rather think the miners commenced operations first, and had it not been for them, the land spoken of would still be a wild morass, peopled only by wandering Spaniards. Neither the Legislature or the courts will ever interfere with the business of mining. It is an industry the Government is fostering. Turn on your hoses miners!"[9]

In the early '70s, particularly in Nevada County, there were numbers of extensive hydraulic claims preparing for development on ground that had remained unworked for years, but with the availability of reasonably priced water, they could be worked quite profitably. Seven or eight years earlier these same claims would barely pay expenses, leaving no margin for profit to the owners.[10] It was noted in a local paper that "... new claims are being opened and old ones are to be reworked. With the new appliances for mining, claims can be worked cheaper and with immense profit to the miner. A few years ago it was thought that all the placer mines were worked out, but it has been practically demonstrated that they have scarcely been touched."[11]

The Orion Mine at Iowa Hill, previously could only wash about five cubic yards with one inch of water, which, with values of three cents per yard, would only equal 15 cents—the actual cost of the water. However, through improved appliances the same inch of water at the same mine, could wash seventeen yards, yielding over fifty cents, representing a fine profit. This logic gave rise to the idea that the standard for judging a good piece of ground should not be the value of the gravel, but rather the power of an inch of water.[12] Thus the term "duty" came into general use meaning the amount of gravel measured in cubic yards that could be washed by a determined amount of water measured in miner's inches. For example, if 150,000 miner's inches (a season's total of 150 days) could wash 600,000 cubic yards of gravel, the duty would be four. The duty for cemented gravel would be much lower, perhaps two. According to one authority, a miner's inch of water, under ordinary circumstances, would wash from one and a quarter to four cubic yards of gravel—average, say three.[13]

The duty of the miner's inch depended on the character of the gravel, the size, grade, nature of the sluices, the height of the bank, the skill of the operator and the size and effective pressure of the column of water. Duty could range all the way from one cubic yard washed per miner's inch to as high as ten cubic yards. From 1870 to 1877, the North Bloomfield mine varied from a low of 3.86 cubic yards to 4.6 cubic yards of gravel per miner's inch.[14]

In addition to an abundant and reasonably priced water supply, another leading factor, contributing to the profitability and rejuvenation of the hydraulic mining industry, was the remarkable advance in the versatility and power of the monitor. As noted in an earlier chapter, the only significant improvement in the design of hydraulic nozzles since the gooseneck monitor, was the rifled nozzle, invented in 1863 by Macy and Martin at Red Dog in Nevada County. The rifling consisted of radial plates of thin iron located inside the cavity of the upper elbow which overcame the tendency of the water to rotate and be discharged from the nozzle in the form of spray. With this new device water was projected as a solid column with no rotary motion.

In 1867 Jenkin W. Richard of Michigan Bluff, Placer County, devised a double-jointed machine hoping to replace of the gooseneck nozzle, but it was soon found to be unreliable and was abandoned as a failure.[15]

The next significant development occurred the same year, at Nevada City, when the Craig brothers (R. R. and J. Craig) introduced the Globe Nozzle into the hydraulic mines. It was designed for the purpose of relieving the amount of water friction passing from the main feeding pipe to the nozzle. Although it represented some improvement, there appears to have been significant difficulty in turning the globe in any required direction when filled with a heavy pressure of water. For this reason it was never brought into general use.[16]

The following year, however, an improved model of the Craig Globe Monitor was designed, using a ball and socket joint as a swivel. Although distinctly better, according to mining engineer A. J. Bowie, under conditions of high pressure it was difficult to operate, sometimes requiring several men to manipulate it. To overcome this problem the Craigs made an improvement to relieve the friction by means of an interior tripod device. Unfortunately, opined engineer Bowie, ". . . after use, these machines sometimes became leaky and difficult to manage."[17]

Despite Bowie's rather unflattering appraisal, the local Nevada City newspaper was quite sanguine when reporting on the new device:

> The North Bloomfield Gravel Company have ordered two globe nozzles of the Messers Craig, which are to be cast at the Nevada Foundry. These are to be thirty inches in diameter inside [the dimension of the globe] and under a hundred and sixty foot pressure, each will be able to pass from 800 to 1,000 inches of water through a seven inch nozzle. Marcelius & Maltman of this city have also ordered one of these huge globes for the Manzanita claims. The amount of water these large ones are capable of discharging under a heavy pressure will be an increase of concentrated hydraulic power beyond anything that has yet been used in deep gravel mining. We also learn that the Craig Brothers have orders from Nevada and Placer counties for eight more of their thirty inch globes. In one claim at Dutch Flat a globe is to be used under pressure of one hundred and fifty inches of water.[18]

The market for the Globe Monitor continued to expand. "This nozzle is gradually coming into use in the hydraulic mines throughout the state," commented the *Gazette*. "An order was received a day or two ago for two of the small sizes, to be sent to Trinity County. They were at once prepared and will be forwarded tomorrow."[19]

However, it appears that perhaps the Craig Globe Monitor did not have an adequate rifle or radial plate installed in the new model as suggested in a later issue. "Whenever the pipe was turned at any considerable angle from the mouth of the globe, the water would lose its force by the twist it acquired when leaving the globe." The article assured its readers that R. R. Craig, the inventor, had corrected the problem by means of a triangular pipe "which straightens the water and prevents it from breaking up immediately after it is freed from the pipe. The new experiment works like a charm, the stream ejected going to the mark in a concentrated body."[20]

During the next few years a number of competing monitors were introduced into the industry, but the Craig model appears to have maintained its market share for a number of years with, at least, moderate success. At the Spanish Hill mine in El Dorado County, where a Monitor was at work under 250 feet of pressure, the owner claimed that the work of two men was saved by its use compared with earlier nozzles.[21]

Finally, in October, 1870, the respected *Mining and Scientific Press* devoted a feature article on the Craig Globe Monitor, reporting that after three years of use the Craig brothers had perfected their invention and it had been used by over one hundred of the most extensive and prominent mines in the state and "has received the highest recommendations of experienced miners." The article then continued:

> The invention consists of a hollow globe or reservoir, in which fits a ball or socket which is provided with an elbow, the globe being cut open on top. The ball revolves entirely around horizontally, and up or down at an angle of about 40 degrees. This play has been found amply sufficient for all ordinary mining purposes and the water, being fed to the discharge pipe from the globe or reservoir, has not the bad effect caused by a short turn in the pipe, like that produced by a common elbow, [the gooseneck] but causes a perfect stream to emerge at any point to which the nozzle may be directed. As a matter of economy, it not only places the water of 7 or 8 ordinary hose pipes under the control of one man, but its durability is so great (one lasting a lifetime) that its extra first cost is seldom noticed, it being in convenience alone worth more to the miner than the difference of cost of canvas hose. No canvas being used, it is not liable to breakage under heavy pressure, and saves the annual outlay for canvas, while the concentration of a larger body of water in one column has been found to nearly treble the amount of execution in comparison with ordinary expenses. The Globe Nozzle and its improvements are covered by three different patents, owned by Messrs Craig, who now have two suits pending in the U.S. Circuit Court against manufacturers and users of alleged infringements thereon.[22]

Perhaps because competing nozzles were beginning to be manufactured by the Nevada Foundry (later to be renamed the Miner's Foundry), in 1870, the Craig brothers moved their operation from that location to the Marysville Foundry in Yuba County. For a time, however, they continued to maintain their headquarters in Nevada City.

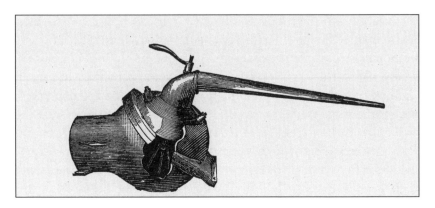

Craig's globe monitor.

During the opening years of the '70s, the demand for hydraulic nozzles capable of operating under the enormous pressures that were being developed at the rejuvenated mines led to a series of new entries into the market, resulting in fierce competition between the fledgling inventors.

In July 1870, the Watson Hydraulic Champion Joint and Nozzle made a brief appearance. Like its predecessors, it was a local product of Nevada County, manufactured by the Nevada Foundry under a licensing agreement.[23]

That same year two other competitors entered the market— Fisher's Hydraulic Chief closely followed by Richard Hoskins' Dictator. Because of similarity in design, the new entries set in motion endless lawsuits and counter-suits over patent infringements which would continue for years.

But in the meantime, despite legal roadblocks, the Craig Monitor remained popular with hydraulic miners for a good number of years. As late as October 1878, the Blue Tent Consolidated Mining Company received a Craig monitor of "monstrous proportions," weighing 1,400 pounds. It came equipped with two nozzles, one nine and the other eight inches in diameter. A newspaper commented that there were but few of equal dimensions operating in the hydraulic mines.[24]

The "Hydraulic Chief," manufactured by Frank H. Fisher, was

also known as the "Knuckle-Joint and Nozzle." It was introduced in late 1870, following Craig's Globe Monitor and Watson's Champion. Although similar to the Gooseneck with reversed elbows, it was connected by a ring with anti-friction rolls. The ring was bolted to a flange which gave the upper elbow a free horizontal movement. Vertical movement was obtained by means of a knuckle-joint located on the outlet side of the upper elbow. This machine was also equipped with vanes or rifles in the discharge pipe to facilitate the flow of water in a straight line. Although it provided a vast improvement in its vertical movement, it also had problems, becoming leaky after use and proved quite expensive to maintain.[25]

A long-sought improvement envisioned by hydraulic miners, in the operation of the powerful new machines, was a device to facilitate their movement in a horizontal direction. Thus an announcement in a local newspaper the following October was read with keen interest. It revealed that Frank Fisher had invented a "regulator" for moving with "ease and celerity" the nozzle of his Knuckle Joint Distributors. It was described as a lever about twelve feet in length, extending from the butt of the nozzle or pipe where it emerges from the distributor. According to the article, the lever was so arranged that it could swing the nozzle around a circle of 50 degrees, stating that a boy of ten years of age could manage it with ease, even when it was under a pressure of 200 feet. Readers were informed that a number of these machines, with the new "guiding device," were in operation at Smartsville, and had received the highest recommendation.[26]

The machine designed and patented by Richard Hoskins, called the "Dictator" substituted an elastic joint in place of the two metal plates used on the Gooseneck. Although simple in construction, this device saw little use. The following year, however, Hoskins improved his Dictator with a new design he renamed the "Little Giant." This model was double-jointed with both a knuckle joint and lateral swivel. The pipe was also equipped with rifles. The Little Giant came with a variety of nozzle sizes, ranging from four to nine inches and this improved machine won immediate and wide acceptance The five and one-half to seven-inch-nozzles appear to have been the most popular. The bearings at the joints were kept in repair by frequent applications of axle-grease or tallow. In operation, all large machines, regardless of the manufacturer, were bolted at the base to a heavy cross-timber and weighted to the ground by a counter-balance, consisting of a long timber projecting to the rear with a box weighted with heavy stones.[27]

Using the time-honored maxim, "if you can't beat 'em join 'em," the Craig Brothers joined forces with Richard Hoskins through an arrangement which permitted both to operate as separate entities. It is very likely Hoskins used this avenue to avoid patent infringements. In later years it was learned that the Craigs and Hoskins had entered into a compact to drive out the competition and control the market — which they soon succeeded in accomplishing. To solidify his market position the shrewd Hoskins, who seemed to possess a superior business sense, obtained the patents on both the old Allenwood Gooseneck and Macy & Martin's radial plates. Both the Globe Monitor and the Little Giant were manufactured at the Empire Foundry in Marysville, which Hoskins subsequently bought out in 1878, acting as both manufacturer and seller. He controlled the market for a number of years manufacturing a variety of sizes which were sold all over the world. Little Giants were available in sizes up to twelve inches with a shipping weight of 2,450 pounds. According to an article in the Marysville *Appeal*, the Empire Foundry even manufactured a monstrous 18-inch Little Giant, the largest in the world, which was delivered to the Blue Tent Hydraulic Mining Company. The same article also mentioned that one of similar size was on order from the North Bloomfield company.[28]

In the meantime, Frank Fisher continued to manufacture his Hydraulic Chief, using the rifle in violation of Hoskins' patent, which resulted in fines, and eventually 30 days in a San Jose jail. He was finally forced out of business and eventually found employment at the mint in San Francisco. Nevertheless, as early as 1870, Frank Fisher had named both the Craig brothers and Hoskins as defendants in a lawsuit claiming patent violations. Over ten years later the suit was finally decided in his favor by a determination of Judge Lorenzo Sawyer of the U.S. Circuit Court. The Little Giant proved

INVENTION AND MATURITY

TO HYDRAULIC MINERS.

I warn Hydraulic Miners not to be misled by the advertisement of Mr. F. H. Fisher. Said Fisher has been beaten in every suit to which he has been a party (respecting Patent Rights on Hydraulic Machines), whether as plaintiff or defendant.

His patrons have already paid over Five Thousand Dollars for license to use machines purchased of the self styled

"Only Reliable Party in the Business Who Protects His Customers."

Fisher's Machine is worthless unless the discharge pipe is furnished with appliances, the patent for which is owned by Messrs. CRAIG, and which Fisher is enjoined from making.

The Machine itself is also an infringement on another patent owned by the same parties.

All persons infringing on either of these patents

WILL BE PROMPTLY PROSECUTED.

I am sole LICENSEE to sell Machines manufactured under Craig's, Rice's, Macey's and Hoskin's patents.

I Guarantee Full Indemnity to all my Customers.

Machines of all sizes always on hand.

If you want a Machine that will give satisfaction, and do not want to pay for it twice over, buy an Improved LITTLE GIANT.

For further particulars apply to

R. HOSKIN, Dutch Flat,
R. R. & J. CRAIG, 304 Montgomery st., S. F.
Or   WILLIAMSON & CORY, Marysville.
Dutch Flat, August 1, 1873.   6v27-2m

to be simply a duplication of the Hydraulic Chief. To make matters even better for Fisher, the Macy & Martin patent, owned by Hoskins, had finally expired. In a rather ironical twist, it is interesting to note that Hoskins also served a short term in jail for refusing to pay a rather insignificant fine for infringing on Fisher's patent.[29]

Earlier, Frank Fisher's "regulator," designed to guide the motion of the pipe in a horizontal direction, was found to be less than a suc-

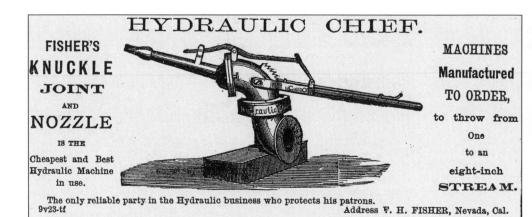

cess and the industry had to wait another five years before an efficient device, to solve this problem, was found. The incident leading to its invention, like so many others, was accidental in nature. David Stokes, foreman at the North Bloomfield mine, while washing a shovel in the stream of a giant, caused it to swerve in the direction opposite to the position of the shovel. He brought this phenomenon to the attention of Henry Perkins, superintendent of the company, who, after a number of experiments, designed a new device known as the "deflector."[30]

The first news of this important invention appeared in the Nevada *Daily Transcript*, on April 18, 1876.

> The Superintendent has invented one of the simplest and most effective improvements extant for controlling the movement of hydraulic nozzles. It consists of a short cylinder about ten inches in length, connected to the end of the nozzle by a flexible joint, and having a lever about four feet in length attached to its upper surface and extending backwards. The operator, when he desires to change the direction of the stream simply moves the lever so as to cause the stream to strike against the side of the cylinder, when the force of the water causes the nozzle to move in the direction required.

While without the aid of this apparatus it requires the power of a man to move the nozzle anyway, with this simple invention the same result can be accomplished by a child.[31]

As a matter of interest, in the *Second Report* of the State Mineralogist, H. G. Hanks, related that the "Little Giant," patented by Richard Hoskins in combination with the "Perkins Deflector," seemed to be the most popular in use at that time.[32]

The Perkins deflector immediately won wide favor with hydraulic mine operators and could be easily retrofitted to existing nozzles. The North Bloomfield Gravel Mining Company named Thom and Allan of the Nevada Foundry as their agents with manufacturing rights to the new invention.

There were immediate look-alike imitations on the market attempting to circumvent the patent rights. It is a matter of some amusement to learn that the Milton Mining & Water Company, with half their stock owned by the North Bloomfield Company, apparently had purchased a number of deflectors from such a source (perhaps in ignorance). In a letter from H. C. Perkins, on North Bloomfield Gravel Mining Company stationary to the Milton Mining & Water Company, dated July 14, 1876, a request was made for the sum of $40 each for all the nozzles with patented deflectors in use by the Milton Co. and to mark each machine with the patent date—"May 16, 1876." Perkins also added, that in the future, all such machines would cost $50 each and that they should be purchased through Agents Thom and Allan of the Nevada Foundry.[33] As a consequence of this rather curt communication from Perkins, Henry Pichoir, writing from Milton corporate headquarters in San Francisco on August 24, 1876, informed his superintentent, V. G. Bell at French Corral ". . . we think also, that if the Perkins Nozzle Patent is of any advantage whatsoever to your washing, $400 for all our claims over the life of the patent, is cheap enough and had better be purchased."[34]

In time, the Perkins Deflector was superseded by the invention of the Hoskins Deflector which differed from the earlier model in two particulars. It was located between the end of the discharge pipe and the nozzle and consisted of a flexible semi-ball joint.[35]

After the invention of the deflector, and as techniques improved, the operation of these high-powered monitors took on a whole new dimension. The operator could, if he preferred, stand beside the machine or sit on the pipe with the deflector lever in his hand carefully guiding its every movement. Saddles were actually designed for this purpose.

In the interests of placing continuity ahead of chronology, the evolution of the "monitor" or "giant," as it became known, should be brought to finality.

The Joshua Hendy Iron Works of San Francisco, in time (after the Sawyer Decision), produced the ultimate monitor and captured the greater part of the market, shipping models all over the world. Although the "Hydraulic Giant," according to August J. Bowie, Jr., was a modification of the Little Giant,[36] the Joshua Hendy Iron Works wisely purchased the rights for the "Double Jointed Giant" from Frank Fisher that was incorporated into the design of their patented "king-bolt" concept. The "Hendy Giant" soon followed.[37]

Sizes ranged from four-inch nozzles to nine-inch in six different sizes weighing from 245 to 1,050 pounds. All sizes could swing horizontally through a full circle and from eleven degrees below to 55 degrees above the horizontal. By the middle 1880s a nine-inch hydraulic giant could discharge in excess of one million gallons an hour or over 25 million gallons per day.[38]

As noted in the Joshua Hendy catalog, "the vertical and horizontal movements are controlled by the use of weighting levers which are usually made on the ground where the giants are in use and therefore are not supplied by the manufacturer." For smaller giants, the size of the timber recommended for the lever was four inches by six inches, twelve feet long; for the larger sizes the timber specified was nine inches by nine inches, 18 feet long. It was also explained that a suitable box or platform should be constructed and weighted by stones to obtain the proper counter-balance. The catalog also noted that: "Deflectors are not part of the giant unless so ordered" and stated that low heads of water did not require deflecting nozzles.[39] According to Joshua Hendy specifications, deflectors were only required with pressures exceeding 100 feet. The company also had sales rights for the

Hoskins deflector and a later model known as the "Smith-La Grange" deflector. The latter was recommended for extremely high pressures exceeding 300 feet.[40] The Hoskins deflector was attached to the spout before the nozzle, and therefore, could permit the use of any size nozzle having a smaller outlet than the diameter of the butt end of the spout being used. The Smith-La Grange deflector was described as being made of cast bronze. This appliance was fitted on to the end of the nozzle instead of on the giant butt. Because of this, a deflector for each size nozzle was required.

Before 1868, in the moderately large hydraulic mines, it was a common practice to use five or six streams of water against a bank and in some cases as many as twelve or 15 undermining the ground. But after the advent of powerful monitors or giants, as they were soon to be called, except for very large operations, only two were used simultaneously—one for cutting and the other for sweeping. Generally a smaller nozzle was used for the cutting operation and a larger for sweeping. For example, it was quite common to use four and one-half-inch and five-inch nozzles together, one cutting and the other sweeping. In small hydraulic mines only one monitor would be used with a change in nozzle size for each operation. In many cases only enough water was available to operate one monitor at a time.

The cutting monitor would play its column of water on the bank in such a manner as to cause the material to cave into the pit. The fall would help break up the gravel. It was then swept into the sluice box by the sweeping monitor. The miners soon learned that one large giant could do more work than two smaller, using the same amount of water. If enough water was available, then two or more giants could work at the same time. At the La Grange Mine in Trinity County, for example, two nine-inch nozzles were used for cutting and two for sweeping. In large mines with plenty of water, even more than four giants could be used at the same time.[41]

In the summer of 1870 the changing conditions in the hydraulic mines was the subject of an editorial in a Nevada City newspaper.

> The general tendency in deep gravel mines throughout the county, is to consolidate — to use more water and a heavier pressure in a less number of

pipes; the size of the latter being increased. Fifteen inch pipes are taking the place of eight, ten and eleven inch. For instance 700 inches of water forced through a six inch nozzle will displace more gravel than the same amount forced through a half a dozen smaller nozzles; and the harder the gravel or cement the greater will be the difference in favor of the larger nozzle. Experience has taught miners that it is economy to use as much water and pressure as possible, and that, where there is one mine that uses too much, there are ten that use too little. If there is any branch of industry in this state that needs to be well done in order to make it profitable, it is in deep placer mining. . . . Good results are already flowing from it, and the result will be a permanent revival in this the most important branch of mining.[42]

At some large operations it was the practice to have two working faces so that one could always be washed during the time the other was temporarily shut down while engaged in moving the pipe or for other reasons. This dual arrangement was accomplished by using a "Y" connection on the distributing pipe with water gates in either direction.[43]

The cutting power of a giant was at its maximum when within range of the unbroken section of the stream. According to one expert, "to strike the stream within this limit, is almost like striking a shaft of steel; to thrust an arm into it is to have the limb torn from its socket."[44]

The force of the stream was so great that 50-pound boulders could be projected a distance with the velocity of a canon ball. Men and animals were known to have been killed by the force of the water at a distance of 200 feet and a powerful man would find it impossible to swing a crowbar through a six-inch jet of water.[45]

According to the *Mining and Scientific Press*, the monitors and methods of application at the Consolidated Mining Company, located at North Columbia, were among the best—if not the best in the state. It advised any tourist "anxious to see the 'Niagara of hydraulic mining' [to] go and see the monitors of the Consolidated mine. . . . It is impossible for one who has not witnessed it to form a correct idea of the effect produced by a 200-foot pressure forcing an enormous torrent of water against these gravel precipices, which seem to actually melt away at its icy touch."[46]

It was important for the "piper," as the operator was called, never to wash point blank at the wall, but rather at a proper angle. It was found that a glancing stream would cut more rapidly than one perpendicular to the bank. A side-cutting action could excavate a far greater quantity of gravel per cubic foot of water used than by "pounding," when the column of water was directed dead on to the bank. The face of the bank was kept square, or nearly so, and the so-called "horseshoe" form was avoided as it was the chief cause of accidents when men in the pit were encircled by the bank.[47] According to some authorities, the best practice was to begin piping at the lowest part of the bedrock and to work across the bank with a tendency to maintain a "nose" of gravel immediately in front of the giant, and to work with a side-cutting action, both left and right of the nose. In this manner the direction of slides would be away from the giant.[48]

Almost from the time hydraulicking came into vogue, accidents, which were all too frequent in the mining industry, were accelerated. The great majority of these were caused by the caving of banks on the hapless miners working in the pits below. In the early period, thanks to the small nozzles and low heads in general use, many were not fatal. James Mason Hutchings, a contemporary writer, has left us a first-hand account of the unique method by which some of these miners were rescued from what, otherwise, could have been a tragic death. "Sometimes when a man has been covered up by the bank falling upon him, not only the stream generally used in the claim, but often the entire contents of the ditch are thus turned on, and with the assistance of every miner who knows of the accident, it is used for sluicing him out, and which is by far the speediest and best method for his deliverance."[49] In many cases, however, the miner was not so lucky. Early in 1865 a Grass Valley newspaper filed this all-to-familiar report: ". . . at about ten o'clock Friday morning, Mr. Richard McDonald was washed down by a rush of earth and water in a mining cut of the Illinois Company at Moore's Flat, and was probably killed instantly."[50] With the introduction of high-pressure nozzles spouting huge volumes of water, such accidents became all the more terrifying. W. W. Kallenberger recalled such an incident that occurred at the Malakoff Diggings.

> I recall the night Al Marten lost his life and Jim Cummins and Gene

Trudell were badly hurt. It was customary to leave a natural abutment against a bank so as to prevent a bad cave of the bank by holding the ground back as much as possible. The night was a wild one. Black, torrential rains, high winds. Jim Cummins was the head piper for the night shift. Marten had the big "giant" and Trudell was utility man. Cummins cautioned Marten against playing the giant against the abutment holding the bank. But Marten took the chance anyway without heed of the order. Then it came. We, safe in our beds a half mile away, air line, felt the shock as if an earthquake was taking place. The roar of the storm without, with wind and heavy rain caused grave misgivings amongst the townsfolk when the shake was felt. Caves were bad enough during daylight hours, but an earth, water-soaked and standing at such great height upon a slippery, clayed base could mean but one thing—tragedy.

A covering bank will create a powerful compression of air and can hurl great chunks of pipe clay and large rocks with great force. Marten was found several feet away from his mud-stuck rubber boots. The force of the air, and aided no doubt by a chunk of clay that must have broken his neck, hurled him completely out of his boots as he ran or tried to run frantically from the caving bank. A huge chunk of clay imprisoned Jim Cummins. But being a powerful man, he managed to heave from underneath before suffocating. Trudell . . . spent months in the French Hospital in San Francisco, with broken bones.[51]

Not all accidents in hydraulic mines were caused by caving banks, however. At the Hathaway mine in Washington Township, the iron lever operating the deflector on a monitor with a three-inch nozzle under 345 feet of pressure broke, causing the monitor to buck. The full force of the water struck the superintendent, Thomas McEachern, who was standing three feet from the nozzle. It hit him full in the breast, hurling him into a ground sluice 20 feet deep. The body flew like a bullet against a pile of rocks in the cut and "two great holes were pierced in his head killing him instantly."[52]

The frequency of accidents from washing down high banks led to the practice of disposing of the gravel in benches. With the heights of banks in excess of 150 feet, the upper half was first attacked and run off followed by the lower. In certain areas it was not unusual for banks to rise 250 to 300 feet and even higher. In later years banks 500 to 600 feet high were not uncommon.

With the advent of new and improved methods and equipment,

the hydraulic mines experienced what amounted to a technological revolution siring a new breed of mining engineer. His expertise not only included the science of geology and mineralogy but a competency in a new discipline. This dealt with the scientific control and application of water in all phases of the mining operation. The new discipline included the construction of dams and the delivery of water through pipes and flumes to the excavation of the ground and the saving of the gold, all through the expedient of water. By a mathematical calculation he could, for example, determine the spouting velocity from a nozzle of any diameter, under any head or column of water pressure, and the amount of water which could flow per second through the orifice.[53]

From experience it was learned that the most effective head range (water pressure) was from 200 to 600 feet. Below 200 feet, the duty (the amount of cubic yards of gravel per inches of water) was too low and above 600 feet, the necessity for heavier iron pipe, anchorage and bracing was too costly. With a 200-foot head, a flow of 2,000 inches could operate a moderately sized hydraulic operation, while a 400-foot head of 1,000 inches could move nearly the same amount.[54] Thus the excavating power of water, in the form of a jet depended more on pressure than on volume. A small high-pressure jet from a three-inch nozzle, for example, under 400 feet of pressure could be used more effectively to undercut a bank and cause caving of a far greater amount than could possibly be washed away. Thus, the speed in which the gravel could be washed into the sluices, after the bank was caved, was the limiting factor. In actual practice, heads of less than 200 feet could only excavate very loose material. On the other hand, jets from four inches to six inches under heads of from 400 feet up could excavate cemented gravel.[55]

As the hydraulic mines were creating a new branch of mining engineering, so, also, they were producing skilled workers with specialties never before known in other forms of mining. In the hydraulic mines an experienced and able pipeman was prized above all others. Mining superintendents and foreman were learning that a good pipeman could work off twice as much gravel as an untrained miner. But the job required not only skill, but the ability to endure the worst kind of working conditions. Typical shifts were twelve hours and, in the winter, pipers often used coal oil lanterns under their full length slickers to keep warm. At the La Grange mine in Trinity County, pipers riding the giants, on occasion, were actually frozen to their machines by the spray from their nozzles.[56] Dennis Stovall has left us a memorable portrait of this unique miner and an incomparable account of the job he performed—a classic in its own right.

> Some superintendents and managers say that a good piper is half a placer mine. To the initiated this statement will not seem far-fetched. Without a good piper who knows a few of the main tricks of handling a giant, an otherwise payable mine will fail. The piper is the fellow who "gets the stuff from the dirt; " and it is the "stuff" that counts. On the Pacific coast, where gigantic hydraulic mines are common, "piping" is an art. "Pipers" are professional men and proud they are of their profession, as they have a right to be, for piping is not picked up in a day. Like prospecting, engineering, and other phases and branches of the mining business, piping is an art that requires years to give perfection. . . . Pipers demand and receive excellent wages, and there are few of them but earn every dollar that their pay-check represents.
>
> To the expert placer-piper the roaring singing monitor that yields to the deflectors slightest touch is as it were, a living breathing thing. A hydraulic giant is to him as a tamed lion to its master—obedient and powerful. None know better than he how to swerve the big nozzle to drive an avalanche of boulders down the gulch ahead of the giant's stream, scattering them like handfuls of bullets shot from a catapult; or to bring that long, deep growl from the monster as it gnaws at the base of the towering red clay bank, till a great slab of a thousand tons topples and falls with a mighty crash from the mountainside. Clad from hat to boots in rubber and wool, the piper is at his post everyday of the mining season, no matter how swiftly the wind may blow or how icily it may bite, or whether the rains pour, or the snows pile the diggings under a mantle of white—he is always there, directing the giant's powerful stream.[57]

After the cutting monitor caved a bank, it was the job of the sweeping monitor to wash the mass into the sluices. This was generally accomplished by means of a bedrock cut, which was a trench carved into the bedrock from the head of the sluices to the working face or gravel bank. These cuts were about the same width as the sluices and often ranged in depth from 20 to 40 feet. Bedrock cuts were also sometimes referred to as ground sluices. If the claim used a bedrock tunnel, the gravel was washed through the cut into its mouth. In many cases the tunnel was located below the bed of the pit and connected by one or more shafts or chimneys, often with a fall of as much as 200 feet. With such an arrangement, when the debris was washed through the cut into the shaft, it had the added advantage of breaking up cemented gravel by the force of the fall. Large monitors, with sufficient water pressure, could wash boulders as large as two or three tons through the sluices. In cases of clay-bound gravel, the operator might wash the material back and forth across the pit bottom one or more times until the mass was free from clay. The term "goosing" was used to express the action of driving or sweeping the debris before the stream, while the term "drawing" meant aiming the stream beyond the debris and cause it to be drawn toward the nozzle.

The bedrock in the pit was usually cleaned by piping. Because gold particles were often lodged into the cracks and crevices of the bedrock, as much as two feet would be cut by the monitor and the material washed into the sluices. If the bedrock was too hard, it was cleaned by means of hand tools designed for the job.[58] During the earlier period, before the introduction of the powerful nozzles, the cement that could not be washed into the sluices was dug up with picks and shovels and wheeled in barrows to the sluice. The removal of the last bit of cement from the bedrock was carried on with great care with picks, knives and scrapers as the richest ground was "what sticks tightest and comes last."[59]

In claims where the grade in the sluice was not sufficient to dispose of boulders and course material, monitors were set up at the lower end to move the material along. In later years the Joshua Hendy Iron Works produced a high-pressure giant they called the "Booster," designed to wash material into the sluices. It was made with no horizontal movement reducing the retarding action of friction and giving the giant maximum velocity. It had the capability of breaking up cement and moving giant boulders.[60]

To aid and, in some cases, to replace the sweeping monitor in driving the gravel through the sluices, a stream of flowing water, known as "bank water" or "by-wash" or "by-water," was directed over the bank, across the pit and into the sluices. In claims where water was very plentiful, bank water was also used for ground sluicing.[61] According to one authority: "Without plenty of by-water to assist in driving the torn down gravel to the sluices, the piper is greatly handicapped, for as much or more of his time must be used in 'driving' as in 'cutting'.... Rather than operate a battery of three giants, it is best to operate only two, using the third for by-water, or to increase the supply of by-water already available."[62]

In later years, at most of the larger mines in the Sierra with banks in the range of 500 to 600 feet, only cutting giants were in general use. According to Charles Haley of the California State Mining Bureau, the material was moved into the sluices by bank water which he referred to as "lead water." On the other hand, in the Trinity and Klamath mines, where the banks were generally much lower, possibly only 30 to 50 feet above the river, booster or "drive" giants were used to sweep the gravel into the sluices.[63]

It is difficult to comprehend today the power and force of water as it moved through the sluices. For a better understanding, it should be noted that the transporting capacity of water is increased when mixed with sand. For example, if one cubic foot of water, which weighs 62.5 pounds moves at ten feet per second, it will have a momentum of 625 pounds. In one cubic foot of material composed of two-thirds water and one-third sand, the weight will be 82.66 pounds, and with the same velocity of ten feet per second, will have a momentum of 826 pounds.[64]

Working in a tunnel next to a roaring sluice could be a very dangerous occupation. At the Polar Star claim near Dutch Flat, two men lost their lives by being washed out of the tunnel by an unusual rush of water, sand and gravel. Their bodies were subsequently recovered at the river in such a mangled and bruised condition they were scarcely recognizable by their friends.[65]

Because of the extreme wear and tear caused by such velocity and momentum, sluices, understandably, were set on as straight a line as possible. The grade was generally six to six and one-half inches per twelve feet (the length of the box). In areas of cement or pipe-clay, a steep grade of perhaps nine to twelve inches per twelve feet was preferred. Sluices varied in width from 36 inches to six feet, and in depth from 30 to 36 inches. Sizes, again, depended on the grade and the type of material passing through.

Riffle plates, because of the beating of cobbles and boulders passing through the sluices, were made from large blocks of wood set vertically about nine inches high and 18 to 20 inches square. If wood was scarce, flat boulders were often used or, in many cases, iron rails. In bedrock tunnels, it was a general practice to use a good part of the bedrock itself to serve as a depository for the gold and amalgam. For example, in the 8,000-foot bedrock tunnel at the North Bloomfield mine, there were 1,800 feet of sluices paved with blocks at the upper end, while the remaining 6,200 feet were left bare.[66]

In mines with a steep grade and plenty of water, the so-called rock sluice was popular. It preferably consisted of a wider than usual box completely paved with cobble stones lapped over each other and inclined downstream. The stones were held firmly in place by nailing strips of board generally five and one-half inches wide on either side of the box, with a cross-piece wedged under the strips at each end. After sand and gravel were allowed to run over the cobbles, it was said that they became as immovable as though set in mortor.[67]

A more sophisticated type of paving for a sluice was known as the Hungarian riffle. There appears to have been several variations in use. The simplest consisted of large blocks of wood protected from wear on the top by means of strap iron. Another was made of angle iron sections, running laterally across the sluice, bolted to iron strips which tied a set of blocks together. Each set was two feet long containing eight angle iron bars, producing a riffle space of three inches.[68] Hungarian riffles were also described as two-inch by four-inch scantlings covered with steel straps. They were often placed in sections, alternately running lengthwise and then crosswise in the boxes. As an alternative, railway iron weighing about 40 pounds per yard, laid across the boxes and spread about two to two and one-half inches apart also became very popular.[69]

In some mines a double set of sluices were placed in the bedrock tunnels so that one could be cleaned up while the other remained in operation. In other claims the topography often compelled the building of branch sluices. Branches were also used where the dump required frequent changes for the discharged tailings.[70]

In most hydraulic mines the sluices were charged with quicksilver to gather up and hold the gold. The general practice was to first run the water through the sluices for half a day before charging, in order to "pack" the sluices. Then the water was turned off and the quicksilver placed in the first 200 to 300 feet of sluices. As a rule from 500 to 600 pounds were used in a line of sluices 5,000 feet long. More quicksilver was required at the upper end than the lower. It was carefully sprinkled on the surface, finding its way into the bed of the sluices and riffles. After piping was resumed, about a flask per day was added.[71] Quicksilver was expensive and, therefore, good operators worked to keep the loss at a minimum. At the North Bloomfield mine, for example, the loss was about eleven percent, however, at some mines as much as 25 percent was lost. Large losses were usually due to old sluices, worn riffles and steep grades. Improper sprinkling of quicksilver into the sluices, in rare cases, could cause a condition called "flouring" where the quicksilver would break up into minute, dull colored drops that were lost. Flouring was usually caused by agitation of the quicksilver or its exposure to air.[72]

The clean-up was customarily set at the time the riffles needed to be either reversed or replaced, and could vary from every few weeks to two or three times a year. Just before the clean-up began the bedrock was washed clean and no material other than water was allowed in the sluices. The wooden blocks were removed in 100-foot sections, leaving one row of blocks to serve as riffles to prevent the gold and quicksilver from passing down the sluice. The blocks were then washed clean and the quicksilver and gold removed with iron scoops and placed in sheet iron buckets. As one section was

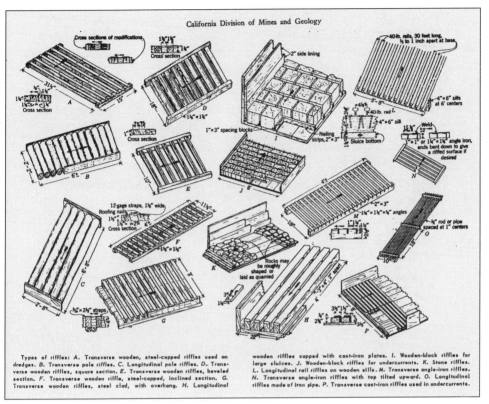

Types of riffles.

cleaned, the residue was washed down to the next set of riffles and the process repeated. During this work a small amount of water was allowed to pass through the sluices to aid in the clean-up process. When this operation was completed, the water was turned completely off and the painstaking job of "crevicing" was begun when holes and cracks in the sluice were cleaned out with brushes and silver spoons to obtain any amalgam remaining.[73]

After the amalgam was retrieved from the sluices, the buckets were skimmed off to remove any foreign substance, then retorted with the quicksilver saved by condensation through a wooden-jacketed pipe and then drained into a bucket of water. The gold remaining in the retort was transferred to a crucible and fused in a furnace. It was then poured into molds as bullion for shipment to a smelter, such as Selby in San Francisco. Larger operations had their own smelter, in which case their gold bars were shipped directly to the mint.[74]

The Polar Star Mine at Dutch Flat had a rather unique method of clean-up because it was one of the few mines that did not use quicksilver in its sluices, but only during the clean-up operation. When beginning, the water was not immediately shut off, but was kept playing on the bedrock until it ran clean through the sluices, then turned off. Three or four men would then commence at the head of the tunnel and pry up the iron rails with an iron bar, leaving the bottom clear. A cleat four inches square, used as a marker, was placed at a distance below in the tunnel. Two hundred inches of water were then released, and all the rails, blocks, ties, etc. were washed in the flow and scrubbed with a small broom. A worker followed the water with a shovel or scoop moving the lighter particles forward, and the gold began to appear. After the gold was gently swept down to the cleat, it was shoveled into a bucket in which 30 pounds of quicksilver had been placed. The sluices below were then cleaned in a similar manner.[75]

Except for periods of clean-up, one man was generally employed to watch the flow of water through the sluices and to prevent jamming caused by tree trunks or boulders. This watchman was also armed with a rifle to protect the sluices from robbers. In some cases a sentry-box was situated on the brow of a hill for a better view by the lookout.[76]

Local newspapers in placer mining areas regularly carried stories of the robbing of sluices, such as: "Two Chinese sluice robbers were killed at French Corral last Friday night. They were cleaning up the Milton Water & Mining Company's sluices when John Moulton, the watchman discovered the rascals, and shot them both to death;"[77] or: "At a late hour on Sunday night last. . . . a person employed to watch Mr. Marcellus' sluices near this city, discovered someone making an attempt to take up the blocks in the sluice boxes. . . . The watchman took as good aim as he could in the darkness and fired. The next morning a dead Chinaman was found a short distance from the sluices on the road leading to Selby Flat.[78]

Small operators would sometimes protect their sluices with booby traps causing a small blast. They were often made with an arrangement of snappers connected to flasks of powder. In one case, it was learned, the result was far from satisfactory as reported in the Nevada *Transcript*. "The miners in Omega have been bothered by sluice robbers so they set up a trap by filling a quicksilver flask half-way with black powder. The remaining space was filled with pebbles and a gun was arranged in the 'stopple.' Upon this a cap was placed so that if someone stepped into the sluice it would discharge the bomb. All the miners except one left the diggings. In time a dog trotted into the sluice and tripped the device. Fortunately both the dog and the man escaped unhurt."[79]

Sluice robbing was so profitable it became almost a science in the manner in which it was conducted. A Nevada City newspaper described the state of the art during its heyday.

> Chinese sluice box robbers have hit upon a new invention for robbing sluice boxes at night. They use a polished copper rod, which they insert between the blocks. The amalgam adheres to the rod as it is withdrawn and is rubbed off into a bag and the operator goes over the box until it is stripped. By this means, a long set of boxes may be robbed in a short time while the water is running, without the blocks or riffles being disturbed, or any evidence left by which the nefarious work is suspected in the morning. Canyon flumes are allowed to run several weeks by their owners without cleaning-up, and these often fail to find enough gold in the boxes to pay the expenses of running, much to their surprise when they should find hundreds of dollars. The infernal ingenuity of these Chinese robbers, which is exercised all over the county, can best be counteracted by a night guard and a liberal use of shot guns.[80]

Chinese were so unfairly treated and disliked by the white population in the California mines, they were undoubtedly blamed for any unsolved robbery that occurred. From the foregoing it would be easy to assume all sluice robberies were conducted by Chinese. This was hardly the case. Many were perpetrated by white men as numerous news accounts attest: "Four miles north of Camptonville one of the partners of a hydraulic mine was inspecting the diggings one night and saw two men in the sluices. The partner returned to the tent to determine if the two men were robbers or his partners. Upon learning they were, indeed, robbers, the six partners pursued the robbers—one turned and fired point blank, killing one of the partners. Unfortunately both robbers escaped unharmed."[81] In this case the robbers were not only white, but murderers as well.

Even with the finest sluice boxes and the best maintained riffles, the very nature of running a mixture of water, sand, gravel and cobbles, propelled by the necessary velocity to carry the mass to the dump, gave certainty to the fact that a portion of the gold would be lost. As early as 1858 it was believed that as much as one-fourth to one-half of the precious metal was washed through the sluices and on to the dumps. A number of authorities were certain that less than two-thirds of the gold was saved by the process of hydraulic washing, the balance passing off with the tailings.[82]

To correct this problem a device known as an "undercurrent" (in early mining literature this word is often hyphenated or used as two words) was designed to be used in conjunction with the main sluice boxes. It was a broad sluice set on a steeper grade, at the side and below the main sluice. Generally it was constructed as a shallow wooden box 20 to 40 feet wide and 40 to 50 feet long, with sides 16 inches in height. Undercurrents could vary in size from 500 to 1,000 square feet and were designed to be eight to ten times the width of the main sluice. The bottom was paved with cobbles, Hungarian riffles or wooden blocks. The grade of the undercurrent was usually one inch to one foot, however, this could vary depending on the type of riffles used.

The material entering the undercurrent was discharged from the main sluice through an opening called a "drop" cut through the bottom at the far end of the sluice, about 15 to 18 inches in width, but could vary considerably. In this opening a grizzly was inserted which consisted of steel bars one inch in diameter and spaced one inch apart, here again, there was no hard and fast rule, for example, at the Gold Run Mine, the bars were four inches wide and spaced from

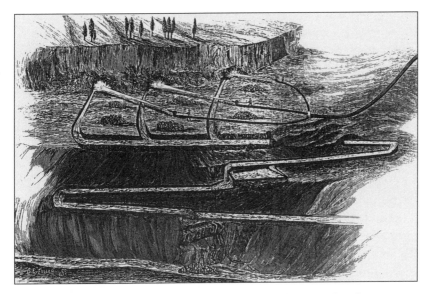

Hydraulic Mining in 1872. A—undercurrent; B—drop; D—distributor.
From Rossiter W. Raymond, Fifth Annual Report, 1873.

three-quarters to one inch apart. The small particles of gravel, sand and gold were then caught up in the riffles of the undercurrent which was also provided with an opening in the lower end to return the bulk of the material received to the main sluice below. Undercurrents could be set in succession, for example, at the American Mine near North San Juan, a row of 20 were installed with such favorable results that by 1876, a total of 52 were in operation. The number used was usually governed by the fall and the distance from the mouth of the bedrock tunnel to the final dump.[83]

The first undercurrent, at least brought to public notice, was designed by R. H. Dunning of Bridgeport Township, Nevada County, and was described in detail, with an illustration in the pages of the *Mining and Scientific Press* in 1861. It appears that Dunning had named Mr. J. Silversmith, the publisher, as his exclusive agent and that all inquiries or purchases should be directed to the

Mining and Scientific Press. [84] In October 1866, it was reported by the Nevada Transcript that Dunning had brought suit against a miner named Jenkin Morgan for patent infringement. The new device was called a "separator" or "under current" and, according to the account, "... is constructed in the bottom of the sluice box and is used to separate the black sand and fine particles of gold from the tailings." The suit was unsuccessful because the defendant proved that the device was only an adaptation of an old and well established principle of the "old Grizzly." The actual undercurrent could not be patented because it was nothing more than a variation of the common sluice box.[85]

Even preceding the undercurrent as a means of retrieving all the gold that could possibly be obtained, the so-called "tail sluice" became a more integral part of the hydraulic mining process and more intensely used. In effect it was nothing more than a long continuation of the main sluice. Generally, it was constructed much wider than the common sluice ranging from 6 to 20 feet. These appliances could be strung out for a distance of several hundred feet to several miles—one actually measuring six miles in length. Tail sluices were often divided into two compartments so that one side could be cleaned up while the other side was kept washing. In many cases tail sluices were constructed outside the boundaries of the claim, and were built and operated by miners who had no ownership in the mine and, thus, only incurred the expense of the sluice. As a result, the reworking of tailings became another important adjunct of hydraulic mining.[86]

In some areas, tail sluicing, when operating outside the claim, was far more than a scavenger-type business. Many became major undertakings involving a large investment, especially in regions where there were a number of claims, both hydraulic and drift, working in close proximity. An April 1860 item in the Sierra Democrat reported: "The Sears Ravine Flume will commence operations next week. The flume is a new work recently completed to take up the tailings of the mining companies, and is expected to be a good speculation. About five or six companies will commence work next week, and soon nine or ten will be contributing to the flume. Chas. Hendel has been in this business in Sears Ravine for years and has made a great deal of money which he is well deserving."[87]

Tailings also became profitable in quite a different fashion as learned through the pages of the Nevada *Daily Gazette*. "In many cases owners of the beds of streams lease them for a handsome sum every year. As the owner is to no expense, this may be said to be the most productive kind of property. These tailings, sent down from hydraulic claims miles above cost nothing, and age and exposure releases the finer particles of gold that could not be caught at the first washing. . . . We know of a number of cases on Deer Creek where the annual filling in of tailings are leased by those owning the ground for from twelve to 1,500 dollars annually, while the expense to them is not a dollar."[88]

The clean-up of undercurrents would occur as soon as it was determined there was danger of wasting quicksilver at the lower ends of the line of sluices. Tail sluices, however, were usually cleaned up only at the end of the season.[89] Undercurrents were charged with quicksilver at the same time as the main sluice and, at some mines, a small amount would be placed in the tail sluices as well. At the North Bloomfield mine, a 24-foot undercurrent required a charge of from 80 to 88 pounds of quicksilver.[90]

From an experiment at the Spring Valley Hydraulic Mine at Cherokee Flat in Butte County, it was determined that the loss of gold through the sluices was even more than most mining men imagined. In 1879, 24 undercurrents were placed along the upper part of the sluices, with one placed at the very end of the tail-race a mile and one-quarter from the head of the flume. At a clean-up six months later, this lone undercurrent was found to have saved between six and seven hundred dollars.[91]

In most cases it was found that 80 percent of the gold was caught in the first 200 feet of the main sluice. However, this percentage could vary considerably. The distribution of gold at the North Bloomfield Mine for a typical year was as follows:

1,800 feet of sluices in bedrock tunnel	92.00%
6,200 feet of bedrock in tunnel	3.75%
300 feet of tail sluices	.95%
Undercurrents	2.50%

At the Gardner's Point Claim, from a total clean-up of $63,000, the yield of three undercurrents was distributed in the following manner: 1st, $1,500; 2nd, $1,000 and the 3rd, $500.[92]

In 1878 at the Manzanita Mine owned by the Milton Mining & Water Company at French Corral, the distribution of gold was:

Main sluice in tunnel, 2,300 feet	61%
Tail sluice, 4,214 feet	30%
Ten undercurrents	9% [93]

Because amalgamation occurs much more readily in warm rather than cold weather, at hydraulic mines located in high altitudes, special precautions were necessary. In fact, quicksilver, in water at 32 degrees above zero, will not amalgamate.[94] The winter of 1880 was so cold the *Transcript* reported that at elevations as low as the You Bet Hydraulic Mine, amalgamation did not take place and a good deal of quicksilver was lost.[95]

Also, in a study of black sand and "rusty gold," it was determined that it was virtually impossible to save all the gold using the hydraulic process, no matter how many precautions were taken. "The tailings of hydraulic and placer mines are an interesting study," State Mineralogist H. G. Hanks opined. "It has been difficult to separate the gold from them in many cases. The difficulty seems to be mechanical, the magnetic residue is so nearly the same specific gravity as the gold that the precious metal is lost by the mechanical force of the water used in concentration. Some of the gold is coated with foreign matter [rusty gold] which prevents amalgamation."[96]

In some mines where extremely fine gold was being washed, copper plates, measuring about three feet by six feet, were placed in the sluices, washed with a solution of nitric acid, to remove all the dirt and grease, and then covered with a coating of quicksilver applied by means of a rag. Using this method, it was possible to recover gold that would otherwise be washed over the riffles and lost. In such operations it was necessary for the sluice boxes to have a very slight grade with the copper plate placed at a nearly level position.[97]

John Hayes Hammond, the noted mining engineer, writing in the Ninth Annual *Report* of the State Mineralogist, was satisfied that the modern placer miner was doing an excellent job in saving the gold from the gravel he extracted, stating: "In the opinion of the writer in most well conducted hydraulic and drift mining operations at least 85 percent and in many cases upwards to 95 percent of the gold tenure of the gravel is saved. Louis Glass, formerly the manager of the Spring Valley Mine is of the opinion that not more than five percent of the gold passed off in the tailings."[98]

In the previous chapter the early methods of bringing water to the hydraulic claims and the use of distributing boxes were described. But with the evolution of giant nozzles, together with enormous water supplies and greatly increased water pressures, a more sophisticated distributing system, at the larger mine sites, was required. From this need the so-called "pressure box" (sometimes called "bulkhead") evolved. It was located at the end of the ditch or flume above the claim. From it the water was delivered into the supply line. Attached to the pressure box, and forming a part of it, was a sandbox, which was situated below the level of the flume or ditch to serve as a filter for catching any gravel or sand carried by the flow of the water. A side gate was constructed at the sand box for the purpose of cleaning it out when needed.

The pressure box was a large rectangular receptacle or cistern constructed of wood and equipped with a grating or grizzly to catch floating debris, such as tree branches or an occasional dead animal. The pressure box was so named because the head (water pressure) was calculated by the vertical distance from that point to the distributing pipe located at the bottom of the pit.[99]

The main supply line descended into the diggings on as straight a course as possible, avoiding angles, rises and depressions. Air valves were usually installed at proper intervals along the line to allow for the escape of trapped air. Where the pipe passed over steep banks it was supported by trestlework. If any angles were necessary along the path of the pipe, they were well braced. The supply line was much larger in diameter at the pressure box, actually assuming a funnel shape, but as it descended down the bank it became uniform in size. The water in the pressure box was kept perfectly still, with no motion before it entered the pipe. This precaution was to prevent the possibility of an air lock which could cause an intermittent stream at the nozzle. In some cases two supply lines were led from the pressure box. The diameter of these pipes, at the point they were attached, was rather enormous in size. For example the pipe connected to the bulkhead at the Selby Hill Mine, near Nevada City, was five feet in diameter.[100]

At the bottom of the claim, the supply line was connected to the distributor pipe by means of a gate. If more than one nozzle was used at the mine site, one or more branch lines would take off from the main supply line. At first, this was accomplished by means of a fork or "Y", but in later years this practice, because of higher water pressures, was found to be unsafe. Gates were then provided for branch lines at a "T" connection. It was considered good practice to use a smaller diameter pipe for the branch lines, in relation to the main supply line, in order to maintain maximum pressure. This could also be accomplished by the proper sizing of the branch nozzles. In hydraulic operations with more modest water pressures, distributing boxes were often used in conjunction with pressure boxes.

At the Southern Cross Mine, the pipe size from the pressure box was 40 inches in diameter, tapering for 500 feet to 22 inches, which was maintained for a distance of 2,800 feet, it then branched into two pipes each sized at 15 inches. At the Malakoff Mine, the pipe at the head was 27 inches tapering to 22 inches, then 15 inches at its branches. Here two nozzles, one nine inches and the other six inches, were operated under a head of 450 feet.[101]

In case of an emergency, it was necessary to have the water turned off at the head. At high-pressure operations it was impossible to safely turn the water off at a gate located at the bottom of the pit. Therefore, a system of signals was established. The piper communicated his desire to have the water turned off at the head to a watchman in charge of signals. At the Polar Star mine at Dutch Flat, the signals were carried out by changing the color of a board 30 or 40 inches square, painted white. Over this a similar sliding board was attached, which could be easily raised or lowered. This motion was communicated by a wire which was led to a convenient spot within easy reach of the piper. The movable board was painted red. The signal was located in a visible place within full view of the pressure box. When the wire was tugged, it was raised, covering the white, which indicated to the watchman, at the head, to turn on the water, while the white meant to turn it off. In later years, at the larger mines, this communication was accomplished by telephone.[102]

On one occasion, a news item, relating to a pressure box, appeared in a mining town newspaper which, in the nineteenth century, might have been considered droll journalism, but today would be better characterized as sick. "A dog fell into the bulkhead [pressure box] which feeds the hydraulic pipe in the Blue Tent Consolidated Company's diggings the other day, and the next place he brought up was against the gravel banks, several hundred feet away. He passed through the pipe a long distance, and made his exit through a five-inch nozzle in a very elongated condition. He left no word how it liked the passage."[103]

From the earliest days of hydraulic mining, one of the many problems confronting the operator was dealing with the numerous boulders which, too often, formed a part of the gravel deposits. With an adequate water pressure and proper sized nozzle, boulders of three or four tons were successfully washed through the sluices. In small operations, boulders too large to go through the sluices were rolled away by hand or dragged by teams in "stone-boats." In some cases a huge boulder was simply left standing. Other methods included breaking them up with heavy sledges or drilling and blasting and washing the fragments into the sluices. However, the most common method of handling boulders was by means of a derrick.[104]

Bedrock derricks were designed with a mast generally about 100 feet high with a boom 90 feet long, set in a cast-iron box placed on sills. The mast was held in position by six guys made of galvanized-iron wire rope one inch in diameter. A whip block equipped with a three-quarter-inch wire rope was used for hoisting tackle. A twelve-foot diameter "hurdy-gurdy" wheel (water wheel) produced the power. Using at least 30 inches of water under a head of about 275 feet, the derrick, under these conditions, was capable of lifting stones of up to eleven tons.[105] Our old friend, Edward E. Matteson, the inventor of hydraulic mining, was also credited, in 1860, as being the first person to devise a successful hydraulic derrick, supplanting the manual turning of a crank. This much-heralded event took place at the Omega Mine, Washington Township in Nevada County.[106]

Ironically, it was at the Omega mine, a few years later, where one of the most horrible accidents involving a derrick ever occurred. While hoisting large boulders with an improperly guyed mast, the entire structure crashed to the ground crushing to death two hapless workmen.[107]

Nevertheless, Matteson's hydraulic derrick appears to have been quite successful. An early model, located at Nevada City, utilized a six-foot water wheel using a pressure of 60 feet with four inches of water. According to a published report ". . . .[the] wheel is of 3-horsepower and can do the work of 15 men turning the crank of an ordinary derrick."[108]

Because derricks were such an integral part of hydraulic mining and required so much man power, improvements were highly prized. At a gravel claim at Michigan Bluff in Placer County, the Van Eman Brothers operated four novel derricks which they considered the best in the state. An article in the *Mining and Scientific Press* gave the mining world this description:

> One man sitting at his ease, attends the derrick alone, turns the water on and off, taking the dirt in and out and dumping the rocks without any assistance, except from rope or pulley. The improvement is the invention of

Mr. D. L. Gorman of this place. It is especially useful where there is a head of water as in hydraulic diggings; for the construction and attachment are such that the force of the water is employed to raise the weights.

The derrick is mounted on a globe, out of which projects a pipe which furnishes the motive power, and by thus mounting it the whole machine can be turned in any direction without deranging the driving power. The pipe carrying the water is discharged into the lower part of the globe, which is stationary. The upper half of the globe is made with a flange, so fitted as to prevent the escape of water. The two hemispheres have an upright post which is fastened at the bottom and extends through the top of the upper section where there is a nut to hold the parts together and yet allows the upper hemisphere to turn independently of the lower. The mast and friction wheel here attached are supported upon the globe in such a manner as to revolve with it.[109]

Waterwheels to power the derricks, the so-called "hurdy-gurdy" wheels, were also improved during this period. The term hurdy-gurdy was used to distinguish this type of wheel from the old fashioned over-shot and undershot wheels in general use. The principle involved in the construction of these new wheels was distinctively of California origin and technically they were "tangential impact wheels."[110] Benoit Faucherie (the same man who built the Magenta flume), patented a turbine wheel with the capacity to operate a 40-stamp mill in 1864.[111] Two years later Leffel's "American Double Turbine Water Wheel" came on the market and, according to one account, was "believed will supersede all others in use."[112] The greatest advance came around 1870 when Samuel N. Knight brought his wheel into production at Sutter Creek in Amador County. It remained the leading water wheel in the gold mines of California until the ultimate waterwheel was invented by Lester Allen Pelton at Camptonville, Yuba County in 1878. The Pelton wheel is still manufactured today, generating hydroelectric power around the world.[113]

During the seasons when water was plentiful, hydraulic mines operated at full capacity 24 hours a day. To illuminate the pit during the nighttime hours, large bon-fires were burned on the bedrock, fueled by pitch wood. Later this method was improved by fashioning baskets made of iron bars to serve as a bed for the pitch wood fires. In this manner, they were made portable to be moved from spot to spot. The use of pitch wood for illumination could became a significant expense. The North Bloomfield Company, for example, spent $8 a night for fuel alone. Kerosene lanterns, with large locomotive reflectors, later replaced pitch wood as the principal means of illumination. However, when generators were developed for electricity, it is not surprising that a hydraulic mine—in this case the Excelsior Water & Gravel Mining Company, at Mooney Flat— would become the first known company in the United States to use electric lighting for industrial purposes. The plant consisted of a 12,000 candlepower Brush Machine which furnished electricity to three, 3,000 candlepower lights placed in prominent positions upon the claim. The lights shed sufficient brilliance to enable the miners to work as readily as during the day. The company calculated that the cost of operation was no more than ten cents per hour.[114] When electric lights were introduced at the North Bloomfield mines, electricity was also generated by a Brush Machine using one-half-inch by twelve-inch carbon rods. It was driven by a four and one-half-foot hurdy-gurdy using six inches of water through a three-quarter-inch nozzle which developed four horsepower. Two lamps were used at the mine-site for a period of ten to twelve hours at a cost of only $2.38 per night. The power that was generated was also used for driving drills and making iron pipe.[115]

As we have learned, the hydraulic mining industry was on the cutting edge of technological advances from the use of wrought iron pipe to electricity, and would keep pace with other improvements and inventions as they became available. In 1875, the Nevada *Transcript*, the leading organ of the hydraulic mines, reported yet another mining innovation. "The Ingersoll Drill, at work in the Milton Tunnel at French Corral, made during the month of April, 90 feet of tunnel. The hand drillers made during the same time, at work on the opposite face of the same rock, only 25-1/2 feet."[116]

The following year, the *Transcript*'s readers were informed: "The Superintendent of the [Blue Tent] Company was connected by the telegraph with the water agent and the office, so that calls from the claims a mile away, for water or anything else needed, is made by telegraph signals, and is almost instantly made known at headquarters."[117]

Two years later, the hydraulic mining industry scored yet another major technological achievement by establishing the first long distance telephone line in the United States. Three of the largest companies in the business, all in Nevada County, jointly organized the Ridge Telephone Company designed to eliminate water losses they were each experiencing along their extensive system of ditches. The new partnership included the Milton Mining & Water Company at French Corral; the Eureka Lake & Yuba Canal Companies, at North San Juan; and the North Bloomfield Gravel Mining Company at North Bloomfield. The California Electrical Company of San Francisco was engaged to install the system which was manufactured by the prestigious Edison Company. The Milton Company, which derived the greatest benefit from the system, bore the major expense, $4,000 of the total cost of $6,000. The system originated at French Corral and from there connected with Sweetland, San Juan, Cherokee, Columbia Hill, Bloomfield, Derbec, Watt, Moore's Flat, Shands, Eureka, and Weaver Lake. A branch line extended to Lake Faucherie via Bowman Lake and another branch connected with the head of the Milton Ditch via South Fork. The total distance was 60 miles. The system included 22 stations with a total of 30 instruments. All of the post offices throughout the system maintained toll offices, and Western Union had a direct line from Nevada City to North San Juan by means of an interline connection with the Ridge Telephone Company. The company maintained a successful operation for over 20 years of service.[118]

At hydraulic mines, where it was virtually impossible, impractical, or unprofitable to run a bedrock tunnel or where the ground did not have sufficient fall, or, in conformance with new requirements following the Sawyer Decision, a device known as a hydraulic elevator served this special need. It was designed to raise the gravel, sand and water out of the pit and into a sluice located at a higher elevation.

The principle of the elevator was simply this—where the velocity of the water flowing up through an orifice is sufficient to cause a vacuum, it establishes a suction through a tail pipe. To accomplish this feat it was necessary to have a head of pressure five time that required to lift water to the same level. The mechanism consisted of a pipe with a throat of smaller diameter at the lower end. A high pressure nozzle at the mouth of the elevator caused a suction which drew material from a sluice through an aperture located below the throat. At the top of the pipe, the fines would fall through a grizzly while the rocks were stacked on a dump. In large operations, three such elevators were put to use at the same time. Two were kept in constant operation, allowing the third to be moved to a new location. It was generally the practice to blast out a sump about four feet deep and 10 or 15 feet square in the bedrock to set the receiving end of the elevator.

The most advanced type of elevator was the invention of George H. Evans patented under the trade-name of "Evans" and sold by both the Risdon Iron Works and Union Iron Works of San Francisco. Its principal features included three suctions—the main suction and two auxiliaries that greatly increased its efficiency. It was available in four sizes, the largest had a throat 20 inches in diameter with a capacity of up to a 10-inch nozzle. With this machine an 18-inch boulder could be carried to a height of 60 feet with a 400-foot head.[119]

In 1879 Colonel W. S. Davis installed an Evans Elevator at Mammoth Bar on the Middle Fork of the American River near Brown's Bar. The water was brought to the elevator through a 15-inch pipe with a head of 400 feet. A three and one-half-inch nozzle was placed in the mouth of the elevator pipe which was twelve inches in diameter and 60 feet long, installed at an angle of 65 degrees. The gravel caught in the strong suction was carried to the top of the pipe where it was discharged and deflected by a heavy one-half-inch steel plate which was hinged at the top of the pipe. The flow of water, sand and gravel was then discharged into the sluices.[120]

After 1884, at the Malakoff Diggings at North Bloomfield, a 20-inch elevator was put into service. It used 1,300 miner's inches of water under a tremendous head of 530 feet, raising 2,400 to 2,500 cubic yards of gravel a vertical height of 91 feet every 24 hours. One of the greatest disadvantages to this type of elevator was its inability to remove boulders measuring in excess of 20 inches, the maximum diameter of the throat.[121]

A more simple type of hydraulic elevator, which did not have this limitation, was known as the Ruble elevator, named after the Ruble Mine in Southern Oregon where it was invented. It was basically an inclined chute laid at an angle of about 17 degrees generally 100 feet in length, including a ten-foot apron which connected the bottom of the chute to the bedrock. The walls of the apron fit closely inside the walls of the main chute. The chute and apron were lined with one-quarter-inch steel sides and three-eights-inch plate on the bottom. It was eight to ten feet wide, with walls that sloped from a height of twelve feet at the bottom to four feet at the top. The first 20 feet of the chute had a solid bottom with the remaining 70 feet, consisting of a grizzly with bars spaced about two and one-half inches apart. Below the grizzly was a false bottom which sloped down from the chute to a sluice box and was placed at right angles and under the main chute. The false bottom and sluice box were lined with light steel, while Hungarian riffles were used in the sluices. In the pit, wings commonly 10 feet high were built on either side of the elevator, one extended almost to the bank. In this way the material could be easily washed to the apron and into and up the elevator. The entire structure was generally mounted on rollers so that it could be easily moved to another location.[122]

At the Redding Creek Mine near Douglas City, Trinity County, a Ruble elevator operated successfully for several years. Water, under a 300-foot head was brought to the pit through a 24-inch pipe, 3,000 feet long. The gravel was cut and swept to near the entrance of the elevator by a giant with a five or six-inch nozzle. A third giant, with a three-inch nozzle, was used from time to time to level the tailings. The elevator was eight feet wide and 60 feet long and elevated the larger material 25 feet. The material over two inches was washed through the elevator by the giant and the undersized through the grizzly into four, twelve-foot boxes, 48 inches wide set at right angles to the chute. Because of the limited height of the tailings pile, the elevator had to be moved three times each season.[123]

The so-called grizzly elevator was very similar to the Ruble elevator except that instead of a solid bottom it consisted completely of iron bars. It was used for the purpose of removing boulders and coarse debris from a mine with insufficient fall to run heavy material through the sluices. The operation of the grizzly elevator required great skill on the part of the piper, as described in an article in the *Mining and Scientific Press*. "Where a grizzly is employed, the lead race brings all the gravel, by-water, dirt and boulders from the diggings to the base of the elevator. . . . it is [then] the duty of the piper to complete the performance. He dexterously separates the boulders from the finer stuff, and juggles them up the incline. Rocks, stumps, roots, and logs alike, find a common dumping-ground beyond the stacker." The grizzly elevator was made from 10 to 15 feet wide and from 24 to 40 feet in length, with sides 8 to 10 feet high. The chute was set at an angle of 18 degrees. While the boulders and coarse debris were washed up the elevator to the dump, the fine material, with the gold, would drop through the bars and be washed through the sluices.[124]

Closely associated with the hydraulic elevator was the so-called inclined sluice. It was employed to elevate material, which had already passed through a long string of sluice-boxes, to a suitable dump. In this manner, using a booster giant, the material could be easily elevated to a height of 35 feet.[125]

The cement mining industry also kept abreast with the times using new techniques and greatly increased magnitudes of its blasts. Instead of using black powder, giant powder, in conjunction with, an electrical ignition system were widely used. At the Blue Point Mine a blast of 50,000 pounds was set off in a main drift 275 feet long with four cross-drifts, each 50 feet apart consisting of a short arm of 80 feet and long arm of 120 feet, (note the change in configuration). A cartridge was placed in a keg in each cross drift making ten in all. These were set off simultaneously by a charge from a galvanic battery. The blast raised 150,000 cubic yards of earth. At another blast, 200,000 cubic yards were raised at the Blue Tent Consolidated Diggings. The entire mass was raised bodily to a height of four feet. The cost of the powder alone, in this blast, was $4,243. The execution was described as "wonderful and the blast a perfect success." At Sucker Flat, 50,000 pounds of Judson Powder loosened a bank 200 feet high, making possible a washing valued at $150,000 in gold.[126]

These great blasts posed little peril from flying debris, because they were detonated so deeply underground. However, there could be a danger from poisonous gas formed from the combustion, which was heavier than air and known to settle into pockets at some distance from the blast. In such cases it could be a hazard to anyone taking a position in such a location. On Christmas eve, 1916, at a blast at Brandy City, Sierra County, after an explosion of 22,000 pounds of powder composed of ten tons of Judson and one ton of Hercules powder, four men lost their lives by asphyxiation.[127]

With the introduction of powerful new nozzles and greatly increased water pressures, many cement mines were no longer dependent on the earlier methods for processing their gravel after the blasting operation. The Little York Hydraulic Mining Company,

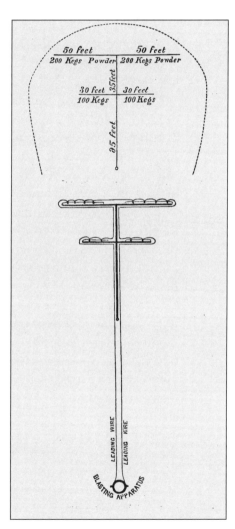

The ground plan and manner of firing a heavy blast in cemented gravel in the 1870s.

owners of one of the most valuable mining properties in California, had always crushed their cement with stamps and saved the gold by amalgamation in the batteries, but, after 1872, with the new appliances, this costly operation was no longer necessary. The mighty giants were now capable of smashing the cemented gravel and propelling it through a long series of sluices, allowing the mills to stand idle. Using but two hydraulic monitors, this progressive, London-based, company was able to wash out $15,000 per month in 1872 and were looking forward to doubling their production in the following year.[128]

Drift mining and hydraulicking were so closely related that, in times of drought, it was not unusual for a hydraulic claim to resort to drifting as a preference to shutting down operations. Similarly, drift mines, after sufficient water was brought to their location, would sometimes shift to the hydraulic process. In a number of claims located in Sierra County, between Pike City and Plum Valley, where each had been worked successfully for years by hydrauckling, they finally switched to drift mining because, "the tenacious clay robs them of too much gold."[129]

In some cases, however, both operations were carried on simultaneously. In the mid-1870s, the Manzanita Mine located in Nevada City, one of the oldest hydraulic mines in Nevada County, successfully maintained a dual operation. During this period, two large monitors, carrying 1,300 inches of water, were constantly in operation washing down banks averaging 130 feet in height, while on the same claim as many as four drifts were being worked. A large force of men were employed picking pipe clay, boring holes and blasting out chunks of gravel. The drifts, which were heavily timbered, were yielding rich pay dirt that was run out in ore cars and dumped into the main sluice. About 50 men were steadily employed in the combined diggings. The drifts produced about $100 to a set of timbers (generally 4 to 6 feet between centers). Altogether, the pay channel was 1,200 feet wide and covered an area of 100 acres of rich company-owned gravel mining ground. In 1874 the combined operations at the mine paid $35,000 to the acre.[130] Similarly, at the Gold

Run Gravels Company, owned by a London syndicate, the principal operations were drifting with the cemented gravel crushed in a 10-stamp mill. However, portions of the mine were simultaneously hydraulicked using two monitors.[131]

To keep abreast with technological advances, hydraulic mining companies began utilizing scientific methods in their selection of sites for future development by employing competent geologists and engineers. The topography of a proposed mine was carefully explored to calculate its length and the depth of the deposit down to bedrock. The course of the channel was then determined and, finally, the value of the gravel. This vital information was ascertained by sinking test shafts over a considerable distance on the channel. The washings from the excavated gravel were used to determine the value per cubic yard. These samplings were averaged over the entire gravel deposit. Two methods were generally followed in these test procedures. For shallow placers, test pits were the preferred choice, but for deeper diggings bore holes were drilled.

At the Malakoff Mine, four prospect shafts were sunk to bedrock and then drifted to get a sampling of the gravel. In this manner, the depth to bedrock was determined and the value of the deposit at that level was calculated by a process of averaging.[132] Under the best of conditions, two and one-half cents per cubic yard was considered the break-even point. Thus, ground that was worth three and one-half cents per cubic yard could yield a profit for a modern hydraulic mine. According to Lindgren, the gravels at the North Bloomfield Mine averaged between 200 and 300 feet deep with the upper 120 feet having little value. The lower 87 feet contained most of the gold. The highest values were within eight feet above bedrock and yielded $1.50 a ton.[133]

The yield from most of the larger hydraulic mines in the state varied from 5 to 30 cents per cubic yard with an average of 15 cents. An acre of ground, in general, contained about 100,000 cubic yards, which yielded a total of from $10,000 to $20,000.[134] However, some ground would have a much higher yield, as noted above, the Manzanita Mine at Nevada City paid $35,000 to the acre.

By the middle and late '70s, the hydraulic mining industry had reached the apex of its development with an investment of over one hundred million dollars and an annual yield of from eleven to 15 million. Although there were still numerous small hydraulic mines operating with six or a dozen men throughout the gold country, the larger companies owned or controlled from one to five miles upon the course of the channel and had invested from $1,000,000 to $3,000,000, requiring the labor of from 75 to 125 men. The magnitude of these operations were both awe-inspiring and terrifying, prompting Thomas Starr King to remark: "All other methods of dealing with the soil for gold are 'one-horse' concerns compared with the hydraulic process. It is fast changing mountains on the face of the state into pits."[135]

CHAPTER 5

The Summit

During the course of this study, we have traced in some detail, the various elements encompassed in the technology of hydraulic mining from its simple and rather crude origin to the state of the art achieved upon reaching maturity. For the purpose of integrating all of the parts into an actual, meaningful mining operation, and, perhaps, to gain a better historical perspective, three of the most successful yet divergent hydraulic mines in California have been selected for a more in-depth view.

The first is the world famous La Grange Mine located in Trinity County in a range of mountains whose waters flow directly into the Pacific Ocean, rather than through the Central Valley. For this reason it was not affected by the Sawyer Decision, and achieved its greatest productivity at a time when most hydraulic mining in the foothills of the Sierra had come to a halt. The second is the Spring Valley Hydraulic Mine, often referred to as the Cherokee Mine, located on Table Mountain in Butte County. It achieved recognition as the world's richest hydraulic mine after overcoming a challenge that became an engineering triumph. Finally, The North Bloomfield Gravel Mining Company and the Malakoff Diggings located on the San Juan Ridge in Nevada County. This mine was enormous in the scope of its operations and the extent of the development work that was accomplished—making it unique in its place in California mining history.

The selection of the La Grange mine as a starting point has somewhat reversed our chronological order, but because its history was so isolated from the events chronicled in the chapter that follows, it was thought best to proceed in this fashion. It is hoped the reader will forgive this transgression. The Malakoff, owned by the North Bloomfield Gravel Mining Company, on the other hand, was so inexorably linked with the central focus of the concluding chapter—its place, as last, was a logical choice.

Although historically diverse, the reader will discover one common theme central to all three, if not to the entire industry—the near heroic efforts undertaken in bringing water to the mine sites as primary and fundamental for any hope of a successful operation. The La Grange mine, unmolested by legislation or litigation, was able to run its course and close for purely economic reasons after achieving a large measure of financial success. The Spring Valley mine at Cherokee Flat very likely went to greater efforts and to more expense in its pains to impound and isolate its slickens than any other similar operation. Nevertheless, mounting costs finally made continuing an impossibility—but not until it had achieved record productivity. The North Bloomfield Gravel Mining Company, unlike the other two, was cut off in its prime with vast amounts of untouched golden gravel deposits awaiting the awesome power of its monitors. Although production was a remarkable $3,500,000, this amount barely covered the preliminary development costs.

The La Grange Hydraulic Mine

Major Pierson B. Reading, former trapper and part-time employee of Captain John Sutter, shortly after the gold discovery in his employer's millrace, reasoned that gold wasn't exclusively confined to the American River, and hurried to prospect the streams near his 26,600-acre ranch San Buenaventura in present-day Shasta County.[1] He proved his judgment was correct when he discovered gold nearby on Clear Creek, a tributary to the Sacramento River. Probing farther west into the mountains, in early July 1848, he made an even richer discovery on the Trinity River. The site, which would later become known as Reading's Bar, was located near present-day Douglas City. In later years, in his reminiscences, Major Reading recalled the events of his gold discovery.

> I prospected for two days and found the bars rich in gold; returned to my home on Cottonwood and in ten days fitted out an expedition for mining purposes.... My party consisted of three white men, one Delaware, one Walla Walla, one Chinook and about sixty Indians from the Sacramento Valley. With this force I worked the bar bearing my name. I had with me one hundred and twenty head of cattle, with an abundant supply of other provisions. After six weeks work parties came in from Oregon, who at once protested against my Indian labor. I then left the stream and returned to my home, where I have since remained in the enjoyment of the tranquil life of a farmer.[2]

Reading was no stranger to the Trinity River, having previously christened it while on a trapping expedition in 1845. He named it "Trinity" because of his mistaken belief that it flowed into the ocean at Trinidad Bay, as erroneously shown on an early Spanish map. A few years later, a short distance downriver, at a location known as Turner's Bar, a company of four Germans and two Danes washed out $32,000 in a single year. One of these miners, a native of Denmark, Peter M. Paulsen, would later play a significant role in the early history of what would become the La Grange mine.[3]

Following Reading's discovery, the bars and gulches along the Trinity River and its tributaries became popular sites for scores of prospectors in their search for gold. Rising above the crowded stream beds, worked by pan and cradle, the surrounding countryside was found unique in its primeval grandeur—revealing a wild magnificence which still lends a kind of mystique to present-day Trinity County. The Scott and Salmon Mountain ranges define the boundary with Siskiyou to the north, while to the east the Trinity Range divides it from Shasta. Towards the south, a portion of the South Fork Range forms the boundary with Humboldt County on the west. This range trends southeasterly through the county, becoming a gentle summit with swells making a natural highway for miles through the southern portion of the county. Between these ranges and those of Humboldt on the west, the entire county is intersected by numerous mountain ranges and steep escarpments forming innumerable valleys through which the Trinity River winds its way.

The site of Trinity County's greatest hydraulic mine was located in Oregon Gulch about four miles above the town of Weaverville. Gold was first discovered there in 1850 by two families from Oregon who were prospecting the area. Although they didn't find an abundance of gold, they did bequeath the name it retains to this day. Extensive mining, however, did not occur until 1851 when it was reported that three men, over a number of weeks, had been washing as much as $300 per day.[4] Mining was restricted entirely to the wet season of the year which usually was all too brief. Among the early miners toiling in Oregon Gulch, James Ward held one of the richest claims, working first with pan and rocker, then with a line of sluices. Over the years he added to his holdings until by 1872 he had expanded his claim to include nearly 120 acres of ground. [5]

Local historian Isaac Cox, in 1858, wrote a rather gloomy sketch of conditions as he found them at Oregon Gulch. "Fronting the river and gulch, two or three bars, extensive enough apparently to admit of employment to scores of miners are worked only, as we are informed, by fifteen or twenty, and yet, when water is in season, good wages are made. What can be done without water? Can any man wonder that our miners leave for parts unknown while our own rich auriferous deposits are treated as though they were undiscovered?"[6] In time, conditions became somewhat better for we are informed by the *Trinity Journal* that by January 21, 1860, there were 100 miners at work earning from six to 20 dollars per day.[7] For a period of a little over a year, beginning April 2, 1860 until August 9, 1861, the place even had its own post office, and the little settlement that formed around it took on the odd sounding name of Messerville. The prosperity in Oregon Gulch was rather fleeting, however, for after the latter date the post office was moved to nearby Junction City. Nevertheless, the first mention of hydraulic mining at Oregon Gulch was credited to James Ward in 1862.[8]

But extensive hydraulic mining came late to Trinity and its neighboring counties of Humbolt, Siskiyou and Del Norte, mostly because of the difficulty in bringing water to the sites. Much of the history of these mines would reflect the great efforts expended in their development. In 1873 Rossiter W. Raymond, Commissioner of Mining Statistics, in his annual report to Congress, quoted B. C. Wattles of Weaverville, giving a good recounting of the state of the mining industry in Trinity County at that period.

> Our mining operations are principally hydraulic. About forty claims are being worked, the most noted of which are the Bolt's Hill, the Holmes, and the Red Hill. The depth of ground varies from 60 to 100 feet, and the average yield of the dirt is probably higher than elsewhere in the state. The product of the county for the past year has been estimated at $1,000,000... In some claims from 75,000 to 100,000 square feet of bed-rock has been stripped; while in others but a small extent has been exposed. We have in this county a range of gravel extending from Trinity Center to the North Fork of Trinity River. This range is about fifty miles in length, with an average width of over five miles, and a probable average depth of 60 feet. The only point at which mining has been prosecuted with vigor has been at Weaver; there has been so far a lack of water elsewhere. Recently a company has been formed and operations commenced to bring water on this range from Stewart's [also spelled Stuart] Fork and another for bringing water from the North Fork to the Holmes and Red Hill ground. . . . The great drawback to the success of our hydraulic mines has been this want of water; and with the construction of these ditches a new era of prosperity will dawn for Trinity County.[9]

In 1872 Peter M. Paulsen, the Dane from Turner's Bar, and a partner, Orange M. Loveridge from Nevada County, began consolidating and developing the Oregon Gulch mines setting in motion a series of events which would ultimately result in the formation of the legendary La Grange Hydraulic Mine. In 1873 the partners formed the Weaverville Ditch & Hydraulic Mining Company which, in exchange for stock, acquired the claims, ditches and water rights of both Paulsen and Loveridge, and also, the mining property of Ward and another miner named Landis. The two former partners received 400 shares each, while Ward and Landis were issued

200 shares. Other investors were also brought into the new corporation generating, including the land, a total capital investment of $600,000.[10] The first order of business was to continue buying up and consolidating the claims in Oregon Gulch and the necessary water rights for a constant and sufficient water supply. However, for their immediate needs two short ditches were constructed from West Weaver Creek. They were four feet wide on the bottom, six and one-half feet on top, and three feet deep, built on a grade of 11.2 feet to the mile.[11]

The Ward mine, absorbed by the new corporation, was the largest and perhaps the richest in Trinity County. Today one can only speculate why Ward would be content to part with such valuable ground for only 200 shares in the new corporation. The face of the claim was 600 feet in length and from 40 to 60 feet in depth. The gravel was intermixed with decomposed quartz and volcanic ash which readily yielded to the action of the water.[12] As reported by the *Trinity Journal*, "Near the top of the divide they are engaged in running a cut; from this they will tunnel a short distance, raise a shaft, and sluice out a hole for a reservoir. . . . From this reservoir it is intended to run the water down a ravine and take it up in a pipe a short distance above the claim. . . . Eleven hundred feet of 15-inch pipe will be required to convey the water on the claim, and 250 feet pressure will be used."[13]

But adequate water continued to be an overriding problem at Oregon Gulch. The drought year of 1875 was particularly severe, allowing only two or three weeks of work for the entire season, yet in a clean-up after only 16 days, the Ward mine produced $3,100. In a subsequent clean-up, later that season, after a run of about 100 hours, the yield was $2,700. It was determined that with a steady water supply the Ward claim could produce $500 per day.[14] But drought was not the only problem that plagued the Weaver Ditch & Hydraulic Mining Company that year. Later in the winter Superintendent O. M. Loveridge complained that the company had suffered a loss by a particularly bad slide on the line of the ditch above the tunnel leading from West Weaver Creek. It required the construction of 15 twelve-foot boxes of flume to repair the gap, but the greatest injury was the loss of precious water while the repairs were being completed.[15]

The two short ditches constructed in 1874 proved to be an ineffective solution to the water supply at Oregon Gulch. The entire season of 1877 was limited to just 303 hours. After the final clean-up of $4,300, the total production amounted to only $8,000 which was calculated to be equivalent to about $25 per hour.[16] The following season, after a particularly wet winter, production proved to be much more satisfying. The ditch was in good condition, running 1,200 inches of water— enough to supply two large monitors. It was also reported that two water-powered derricks were in operation handling large boulders.[17] As late as June, the Ward claim was able to work about five hours a day. That same month, after a clean-up of $5,300, the total for the year was raised to a gratifying $14,000 and there still remained enough water for at least 30 more days of washing. After the final clean-up for the season, which produced $4,700, the total had reached a new high of $19,000.[18]

Realizing that much more capital investment would be needed for proper development, the Weaverville Ditch & Hydraulic Mining Company in 1879, sold its holdings to a group of local Weaverville stockholders. The new company was incorporated as the Trinity Gold Mining Company on September 23 with 100,000 shares of capital stock at a par value of $20 per share. The new president, Dr. J. M. Selfridge, purchased all but 20 shares, allowing the four new directors—Emerson, McWattie, Little and Knight a token subscription of five shares each.[19]

According to the Eighth Annual Report of the State Mineralogist, the most valuable ground owned by the new corporation was still the Ward claim. The channel ran almost due east and west and extended from Oregon Gulch to Weaver Basin. On the summit of the mountain the gravel belt was about one and one-half miles wide. The elevation was 3,100 feet with the belt tapering down to the foot of the mountain, a distance of three-fourths-of-a-mile. The gravel was estimated as being from two hundred to four hundred feet deep.[20] But, not surprisingly, the water supply was still inadequate to properly work the mine. As late as the season of 1888, the Ward mine was restricted to just 369 hours for a total production of a scant $8,000 which translated into $21.28 per hour. Runs were not measured in months or weeks or even days, but rather in hours, running from as few as 100 to 376.[21]

By 1890, when available, 1,500 inches of water were used with heads of 360 feet and 280 feet through three different monitors with nozzles of five and six inches. The sluices were 260 feet long, eight feet wide and three feet deep. They were paved with wooden blocks twelve inches square. Only one undercurrent was in operation measuring 12 by 48 feet, again with block riffles, six inches thick. At that time the average daily run was just two and one-half hours. It was reported that 90 percent of the gold saved was found in the first eight boxes. The tailings were dumped into Oregon Gulch and from there they flowed into the Trinity River above Junction City.[22]

Despite some nominal improvements, the Trinity Gold Mining Company had done nothing in the way of development work to increase the chronic water shortage.

In 1893 the Trinity Gold Mining Company sold their holdings to a syndicate of Colorado and French capitalists for $250,000. The Baron and Baroness Ernest De LaGrange had arrived in the United States in the latter part of 1892 with an eye on mining investments. While in Denver the Baron became acquainted with L. L. Bailey of the firm of Gelder, Bailey & Company, owners of large holdings of mining properties in both the United States and Mexico. Through Bailey, the Baron learned of the opportunities still available in hydraulic mining in Trinity County, an area exempt from the Sawyer Decision because the Trinity River flowed into the Pacific rather than the Sacramento Valley. In 1893, Bailey, together with the Baron and Baroness, arrived in Trinity County and were immediately impressed with the possibilities the mining ground at Oregon Gulch offered if properly developed. They found that the Oregon Mining District was a glacial deposit which filled the ancient channel through the Oregon Mountains. The district began

The tailings sluice at the La Grange Mine. Note the road to
Junction City actually runs under the flume.
Courtesy, Trinity County Historical Society Collection.

in Oregon Gulch on the west side of Oregon Mountain, with the McCarty and Dyer locations showing a depth of gravel of about 50 feet then following up the gulch through the Ward Placer to the Oregon Placer. Crossing the mountain divide, the gravel gradually increased in thickness to a depth of five to six hundred feet. [23] At the time of purchase, the mine comprised five major claims: the McCarthy, Dyer, James Ward, Oregon Gulch and the Loveridge. According to an article in the Tenth *Report* (1890), the area of these claims were as follows:

McCarthy Claim	39.04 acres
Dyer Claim	4.00
James Ward Placer	119.73
Oregan Gulch Placer	159.43
Loveridge Placer	<u>109.40</u>
Total	432.30 [431.60] acres [24]

In early April the transaction was completed and by the 22nd the new company was incorporated under the laws of the State of Colorado, capitalized at $5,000,000. The officers of the newly formed organization included: Baron LaGrange, President; E. Saladin, General Manager; William H. Radford, Superintendent; H. Duvergey, Secretary; and J. B. Doyle, Foreman.[25]

In addition to the La Grange Hydraulic Mine, French capital would subsequently be invested in the *Campagnie Francais de Placers Hydraulique de* Junction City and the Minersville Hydraulic Gold Mining Company. In all, about $1,000,000 of French money was invested over a period of three years.[26] It is worthy of note that the former property, located one-half mile west of Junction City became a quite valuable hydraulic mine. By 1894 it controlled 1,300 acres of auriferous ground 200 feet above the river. Two pits were in operation, with banks of 150 feet and 40 feet respectively. The entire height was gold-bearing even to the grass roots. As many as six giants were in operation in the larger pit with nozzles of six and seven inches. The other pit operated with one large, eight-inch nozzle under 100 feet of pressure. Water was obtained through an eleven-mile ditch taking its water from Canyon Creek. In addition there was another smaller two-mile ditch beginning at Mill Creek. The face of the banks were 2,200 and 1,000 feet long. Although the gravel was free with no clay and few boulders, bank blasting was practiced to increase the duty which was only two cubic yards owing to the limited water pressure. About 900 pounds of powder were used each month. The gravel was washed through two bedrock cuts four and one-half and five feet wide with lengths of 300 and 900 feet. The first 250 feet of flume were cleaned up every 15 days. The mine operated day and night illuminated by electricity. Telephones had also been installed. Mr. F. Heurtevant, who made his home in Junction City, was general manager.[27]

The development work at the La Grange mine began with amazing speed. The MacLean Brothers of Denver, an experienced firm of mining contractors, commenced the pressing job of bringing water to the site from high in the mountains at Stuart's Fork.

Although 300 men were employed for this operation, the contractors complained that they were hampered by a lack of proper tools such as large plows and scrapers, also they found a scarcity of work animals. It is interesting to note the changes that had taken place in construction methods and materials at this later date. In place of cast iron—steel pipe was brought to the site from a firm in Springfield, Illinois—the freight alone costing as much as $9,000 for a single shipment. But the job was not easy. During the construction, the rugged terrain and high elevations led to both landslides and snowslides, not only delaying the progress of the work, but, more importantly, costing the lives of numbers of workmen.[28]

At the mine site that first year, the new company used as much as three tons of powder to bring down the banks for want of water. It was also necessary to haul lumber a distance of four miles for the construction of sluices and riffles. Three monitors were in use, two were very large with 15-inch nozzles, the other was a more modest seven inches. Two derricks were also in constant operation. The mine still only utilized one undercurrent again described as measuring 18 by 48 feet, but divided into three compartments. At that time 37 men were employed, and Mr. W. H. Radford of Weaverville had been retained from the Trinity Gold Mining Company as superintendent.[29]

Although the water system would not be completed until 1898, the *Mining and Scientific Press* noted that as early as the Spring of 1894, water was running through the La Grange ditch, with the exception of a section of about three miles at the head, which was still covered with snow and where some slides had not been cleared. About 800 inches of water were being delivered even though Coal Creek water had not yet been turned on. It was anticipated that this addition would increase the supply by 300 or 400 inches. The article concluded, "the mine has a better water supply than ever before."[30] The following year an update on the progress of the ditch noted that 20 men were then at work on a two-mile tunnel with work being prosecuted at both ends. A shaft had also been driven in the middle, making four faces available for working at the same time. The tunnel was designed to increase the volume of water by connecting Rush Creek and Stuart's Fork which would save nine miles of ditch digging. Its construction had been let to Smith & Ringwood, a subcontracting company, who had already completed a smaller tunnel from the West Weaver, measuring 462 feet long, six feet high and five and one-half feet wide. The excavation was begun on September 17, and finished December 5th.[31]

Water was first turned into the system in April 1898, requiring a full nine hours to reach the mine site. To safeguard the system, ten ditch-tender's cabins were constructed along the course of the channel spaced four miles apart. Each was connected by telephone and stocked with six months provisions.[32]

Commenting on the progress of the development work, the Thirteenth Annual *Report* (1896), stated there was then a new source of water from Stuart's Fork, making available 6,000 inches which would make work at the mine practically continuous. The new ditch, which was 12 to 15 miles long, required a 8,394-foot tunnel. The article added that the entire distance would be flumed six feet wide on top and two feet on the bottom, and four feet deep. Great progress had also been made at the claim. It was noted that five giants with four-inch to seven and one-half-inch nozzles were in operation (the 15-inch nozzles were never mentioned again) along with three derricks with 50-foot, 81-foot and 90-foot booms. The "Loveridge" derrick (a uniquely rigged device) moved rocks weighing four tons, making 24 trips per hour. The pits were lighted by four arc lights of 2,000 candlepower furnished by a dynamo of eight and one-half kilowatts driven by a three-foot Pelton Wheel. All parts of the works and ditches were connected by telephone, 20 miles in extent.[33]

Although much of the system was in operation by early 1898, improvements and additions continued into the next century under the able direction of a French Engineer, Chaumont Quitry. The principal storage site for water was at a dam on Lower Stuart's Fork, high in the Trinity Alps. From there it flowed through a ditch about seven miles long until joining Deer Creek. Here was located a diversion dam which led to a flume connecting a system, consisting of 29 miles of ditches, tunnels, flumes and three inverted siphons, made with steel pipe 26 inches and larger. The deepest was located at a site called Bridge Camp which carried water through a 30-inch steel pipe, one-half-inch thick with a depression of 1,041 feet creating a pressure of 452 psi, making it the greatest inverted siphon in the world. It was anchored at all angles and buried for most of its length in a trench three feet deep and three feet wide. There were eight, three-inch valves on the inlet side and five manholes at points about 1,000 feet apart. Expansion joints were provided at each end of the pipe. At its lowest point there was a four-inch gate valve which served as a blow-off cock to be used at any time it was necessary to empty the pipe. At the stream crossing, it was supported by a wooden truss bridge, twelve feet wide and 138 feet long. Ascending on the opposite side, the water was carried a mile and a half through a flume to the tunnel, described earlier, driven through the divide that separated Stuart's Fork and Rush Creek. Here, at the head of Rush Creek, was the last leg of the system leading into a reservoir at Oregon Mountain, 650 feet above the mine with a storage capacity of 3,000 miner's inches. A penstock situated above the mine led the water into the pit. The overall cost of the water system was no less than $450,000, with the expense of the flumes alone, running from $5,000 to $7,000 per mile. The main water system consisted of 23.85 miles of flume and ditch, including the three inverted siphons, aggregating 8,754 feet in length, and eight tunnels totaling 11,134 feet, making an entire length of 27.61 miles of mining ditch.[34] In 1897 the *Mining and Scientific Press* gave their readers another glimpse of the latest events at the La Grange mine:

> It is estimated that the La Grange people have spent or will spend over $1,000,000 in improvements at the La Grange mine. The newer work such as the flume, inverted siphon and tunnel, cost close to $500,000. The new flume is 11 miles in length requiring 3,000,000 feet of lumber. Three giants operate at the same time. Steel cables are swung across the diggings, and these are attached to travelers fitted with platforms which are lowered to the bed, loaded with boulders and are then hoisted and moved away [the Loveridge Derrick]—all work is done by water power.[35]

High on a hillside overlooking the broad expanse of his mining property, then consisting of over 3,200 acres of gravel deposits, Baron LaGrange built a beautiful chateau as a home for his family and a place to entertain his many guests. Constructed at an initial cost of $65,000, over the years, as the mine became more profitable, tens of thousands of dollars were invested on the interior and furnishings. Stained glass windows, tapestries, statuary, rare art, antiques, all mostly imported from France, made this home, locally called the castle, a showplace for all of Trinity County. Even two Persian cats were imported for added embellishment. The top floor of the three-story structure was used by the Baron as his personal office. The focal point of the second story was a large rectangular-shaped dining hall, surrounded by a number of guest rooms where European nobility and distinguished guests from all over the world were royally entertained. The lower floor was reserved for a huge kitchen, dear to the heart of any true Frenchman, boasting an oven large enough to accommodate a whole pig. This floor was presided over by a highly sought-after French Chef, E. Anssabel.

On the other side of the hill, not visible from the castle, the mine also supported what amounted to a small town complete with a boarding house, bunk house, miner's homes and a school for the children of resident employees. Also, two large barns had been constructed to shelter a herd of work horses along with their feed and equipment. In addition there were a blacksmith and machine shop, compressed air plant, ice plant and electric generator plant. The La Grange mine also owned and operated two sawmills to supply the large amounts of lumber required for flumes, riffle blocks and other needs.[36]

Accounts vary with regard to the number and size of monitors in operation at the pit, but it would appear even in times of drought, two were kept busy from 21 to 24 hours a day. During times of ample water, three or even four were often in use. According to an article written for the Trinity County Historical Society Yearbook, "Two giants work together cutting down the bank called the 'cave.' A smaller giant known as Little Joe is located at the bottom of the pit to

"Departed Grandeur'—Although showing years of neglect, the "Castle" at the La Grange Mine still reflects a bit of its former elegance.
Courtesy, Trinity County Historical Society Collection.

sweep material into the sluices," although no date was indicated, it would appear to be from a rather early period, because "little Joe" could hardly power the huge boulders through the sluices as was customary in later years.[37] Although a Joshua Hendy Catalogue, published in 1922, remarking on the power of the Hendy Booster Giant, disclosed that at the La Grange mine, boulders from two to four tons were washed through the sluices, a United States Geological Survey Bulletin published in 1909 made the astounding claim that boulders as large as seven tons were washed through the flumes at the La Grange mine.[38] From a less creditable publication one would be tempted to dismiss this statistic as erroneous. Seven tons was an amount considered as a huge weight even for a jumbo derrick.

An information circular published by the United States Bureau of Mines reported that, at a later period, when the mine was visited by their engineers, four nine-inch nozzles were in operation, two were used for cutting and two for sweeping.[39] A report in the *Mining and Scientific Press* noted that, in 1903, 25 men worked the La Grange mine with three giants in operation using 3,200 miner's inches of water. Other accounts mention that as many as 30 men were working at the mine.[40] The number of men employed and the number of giants in operation, of course, depended on the season and the amount of water available. For example, even with the outstanding water system at the La Grange mine, during a particularly dry year in early 1912, only two giants were in operation and even these had to be shut down at intervals to let the reservoirs fill.[41] Hydraulic mines were constantly being improved as new techniques and equipment were introduced, making it necessary for the historian to date varying accounts with as much accuracy as possible.

At the turn of the century, the mine had a top sluice or upper sluice, 108 feet long, six feet wide, and three feet deep, with a grade of eight inches to every twelve feet. It was paved with block riffles of three different sizes ranging from eleven by 13 inches to 16 by 16-1/2 inches. This section was followed by a series of bedrock ground sluices, 100 feet long. Below these cuts were 43, twelve-foot boxes, paved with block and stone riffles. At the end of the main sluice was a chute with right and left undercurrents. One was 18 feet wide by 38 feet long, while the other was 24 by 46 feet.[42] At a later date there appears to have been some changes in the sluices. The sluice-way then reached a length of 2,400 feet, constructed of boxes four feet high, six feet wide set in bedrock. The sluices were lined with 40-pound steel rails set on the bottom. Each section contained about 30 rails (140 sections). One thousand cubic yards of material was washed through the sluice-way each hour.[43] All accounts agree that the dumping ground extended to Oregon Gulch for a distance of four and one-half miles toward Junction City. The sluiceway ended on a high trestle which dumped tailings into the gulch. The canyon had a fall 200 to 300 feet per mile opening up on the Trinity River. In order to secure a right-of-way, the company purchased all the land situated between the river and the mine. What was once the village of Oregon Gulch soon became covered with debris 20 to 200 feet deep. Only the church was saved which, in an heroic effort, was moved to Junction City.[44]

The channel was later described as one-half-mile wide and tapered to the top of the mountain. The bedrock sloped at an angle of 70 degrees on the north side and 54 degrees to the south. The gravel ranged from a depth of 100 to 400 feet. As in many auriferous gravel deposits, the richest was found in a layer of blue gravel about 15 feet thick which, combined with another 15-foot layer of red gravel and cemented gravel, comprised the pay dirt. The yield was a very satisfying 18 cents per cubic yard.[45]

The La Grange mine began its years of greatest productivity 15 to 20 years after the Sawyer Decision and, therefore, enjoyed the advantage of nearly fifty years of accumulated technology in hydraulic mining. As a result, new and better techniques were introduced at the La Grange mine never before used by its predecessors. For example, it appears the traditional pressure box was not used. As the *Mining and Scientific Press* reported: "At the top of Table Mountain, a reservoir is supplied by the canal system, from which pipes radiate to the various giants, 15, 18, and 22 inches in diameter and under various heads." This description would indicate that the lines did not all lead to the same pit or the heads would be identical. Ten years later the same publication informed its readers: "From the penstocks are three main pipelines carrying water to six giants working under 450 to 650 feet of head [quite a variation]. Three giants, and one smaller one work at the same time. The largest pipe is 30 inches, the smallest 15 inches."[46]

While on a stay in Europe in 1899, the Baron lost his life in a tragic drowning accident at his estate in France. His father, the Baron Alexis De LaGrange, succeeded as president of the company, reportedly, with very little change in the tenor of operations. Upon the father's death in April 1905, however, the mine was sold to a group of eastern capitalists (referred to by the State Mining Bureau as a French syndicate)[47] who had the good sense to hire an outstanding mining engineer, Pierre Bovery, as general manager and C. Jensen as foreman. The "castle" was occupied by Bovery who continued the grand life-style of his predecessors in a manner that would have gained their full approval; even adding to the furnishings and interior decor.

For the first few years after the La Grange Company took over operations, the annual mining production was about 2,000,000 cubic yards of material. This figure was gradually increased until the mine had a production of 5,000,000 cubic yards of gravel annually. Under the new management, 29 miles of decaying flumes were rebuilt and water storage facilities were increased by the construction of three new reservoirs. At that time the mine was producing over $300,000 per year.[48] Between 1899 to 1911 the cost of water had been reduced to the rather incredible figure of 1/32 of a cent per miner's inch if this figure is to be believed. Bovery also purchased seven miles of 30-inch pipeline from the defunct Sweepstakes Mining Company and added 13 miles of additional ditches. By 1911 the complete water system at the La Grange had risen to a cost of $702,700, as itemized below:

1. 640 Acres of Land for Reservoirs	$ 3,200
2. Dam Lower Lake	15,000
3. Stuart's Fork Siphon	45,000
4. Stuart's Fork Flume	90,000
5. Rush Creek Tunnel	94,000
6. Chaumont Quitry Division	165,000
7. Weaver Basin Siphons (3)	56,000
8. Oregon Mountain Reservoirs	9,000
9. Lateral Ditches, 14 Miles	10,500
10. Sweepstakes Company Pipeline	215,000
	$702,700 [49]

Pierre Bovery, it appears, was responsible for the complete changeover of the riffles in the sluices from wooden blocks and cobbles, which required replacement, every two to three weeks, to steel rails. By 1911 there were 3,000 feet of sluiceway riffled solely with six-inch rails. It was reported that in that year 300 tons of steel had been

La Grange Mine. Note the steel rails in the sluice.
Courtesy, Trinity County Historical Society Collection.

received at the mine to replace worn-out riffles. The previous year 150 tons had been required.[50] The 40-pound steel rails were all laid crosswise except for a few lengths near the upper end of the box which were laid lengthwise. This configuration was used to help give the material a faster start. Part of these rails were set eight inches apart and lasted only two months. Another set of lengthwise rails were set five inches apart and lasted four months. The crosswise rails, which were all set five inches apart lasted a period of six months. When the rails were too worn at the top they were removed, heated in a special furnace and straightened. This recycled material was then used to line the boxes. The rails were shipped from San Francisco already drilled and cut into six-foot lengths. In the sluiceway, about 1,400 feet from the head, the material could be diverted into an alternate flume by means of a steel door which led to another part of the dumping ground. This feature gave more area for dumping and facilitated the clean-up of the lower sluice.

Bovery limited his clean-ups, for the first 40 or 50 sluice boxes, to just three times a year, while the lower boxes were cleaned up even less often. Below the forks of the sluices, controlled by the iron door, the process was only carried out every other year. He was just as precise in calculating the amount of quicksilver used and its scheduling. In the sluiceway, one pint was sprinkled into the first 30 boxes every 36 hours, while one quart was used in boxes 31 to 100, every two weeks. From that point to the end of the flume, one quart was used every two months. It was determined that only 112 pounds of quicksilver were actually lost per year, and this, mainly through flouring. Because of the efficiency of these procedures, the cost of production was only two cents per cubic yard. The whole operation was so well mechanized, Bovery only required 33 employees to run the entire mine, even including ten ditch-tenders. An incredibly small labor-force for the size of the mine. Wages were $3.00 per day for ten hours work. Pipers and their assistants received a higher wage but worked a twelve-hour day.[51]

Every ditch tender's cabin was fitted with an electric call bell attached to a float in the flumes. In this manner, if the water level was either significantly raised or lowered it would sound an alarm.

The water supply was further improved during this period by laying a new pipeline from Stuart's Fork. At the same time considerable repairs over the entire system were implemented.[52] Later in the year, on December 20, 1913, according to an item in the *Mining and Scientific Press*, a major problem developed at the mine when the inverted siphon at Stuart's Fork blew up. Apparently the pipe had been replaced at some time after its initial installation, because the report mentioned that it was 36 inches in diameter (the original was 30 inches). The failure occurred in the section that carried water across the canyon. The account continued: "A wagon and 22 horses and 11 men have been trying to transport a new steel plate to the break in the treacherous steep ravine, but without success." Finally the difficult task was completed by transporting the steel over Buckeye Ditch. An expenditure of $2,000 was necessary just to make the delivery. Complete repairs were accomplished the following month and, with an abundant water supply, hydraulicking continued day and night with returns amounting to over $1,000 per day.[53]

During Bovery's tenure at the La Grange mine he was constantly experimenting with new ideas, many adding substantially to the technology of the industry. For years pipemen had worked under the threat of a broken kingbolt which could cause the elbows to separate at the joint on the giant (Hendy Giant), particularly when operating with extremely high pressures—leading to almost certain death. To prevent such an accident from occurring, Bovery successfully designed a safety clutch and equipped all of his giants with this protective device. For the comfort of the pipeman, he also designed saddles which were installed on the giant's barrels.[54] For the purpose of regulating the flow of water from the mine's reservoirs, regardless of the water level, Bovery installed automatic "floaters" which maintained a constant volume. For maximum efficiency when piping, it was desirable to place the giant as close to the bank as possible, but this practice posed an extreme danger to the pipeman and his assistant. In an effort to avoid such a problem, Bovery perfected a method of operating the giant remotely by means of electromagnets. But because of the danger of short-circuiting from the spray, which could cause the giant to spin in the opposite direction, he abandoned the idea. However, he later experimented with compressed air using remotely placed electrically operated valves, but this idea was never implemented.[55]

It was the practice, during the time Bovery was superintendent, to undercut the 600-foot bank along the bottom, causing the material to fall a considerable distance. In this fashion much of the cemented gravel was broken up in the fall. The masses that did not become crushed small enough were then blasted. Boulders and pipe clay were drilled with an Ingersoll wood-boring machine with a seven-eighths-inch bit and then blasted. Power was furnished by a small water-driven air-compressing plant. By this time derricks were seldom used to handle boulders at the La Grange mine. The high-pressure pipe was sufficient to clean the bedrock — no scraping was necessary.

In order to better deal with the extremely high water pressures developed at the mine, Superintendent Bovery modified the deflectors on his giants in a manner referred to, in a Geological Survey bulletin, as "bootleg". This new device became so successful that it was later manufactured and sold by the Joshua Hendy Iron Works as the La Grange Deflector.[56]

With the outbreak of World War 1 in Europe, Pierre Bovery, a loyal Frenchman, returned to France to fight for his country. He even attempted to entice some of his fellow countrymen, who worked at the mine, to join him. How successful he was in this endeavor is unknown. At about the same time, in 1914, The American Gold Field Company Ltd. acquired a major interest in the mine and Louis Webb succeeded Bovery as manager. Because of skyrocketing prices brought on by the war, the mine was soon operating at a loss. To add to the gloom, on the night of January 17, 1917, the machine shop was completely destroyed by fire. Conditions got no better when a landslide broke a large section of flume and freezing weather further hampered the water supply. In August 1918 the mine was forced to close with much of the equipment sold for scrap.[57] Charles Scott Haley commented that "a large quantity of low grade gravel still remains but it would be necessary to construct a rather costly cut or tunnel to open the mine."[58] It, thus, lay dormant in a state of disintegration until 1927.

In that year it was sold to a new corporation known as La Grange Placers, Inc. This group was headed up by Dr. Byron Stookey a noted New York brain surgeon. His brother, Dr. Lyman Stookey, professor of biochemistry at the University of Southern California Medical School was sent north to manage the property. Under his administration, efforts were made to repair a section of the Rush Creek portion of the ditch, but it proved a hopeless endeavor and was never completed. To be a profitable operation the mine not only required the costly new bedrock cut, mentioned in Haley's report, but the neglected water system needed to be completely renovated. The new owners apparently were not prepared to make the huge outlays of money required to place the mine in working order, it thus lapsed into a state of prolonged neglect.[59]

In 1935 the State of California constructed a highway from Weaverville to Junction City, a total distance of ten miles, crossing the main pit on Oregon Mountain. The state used monitors from the closed mine to cut the roadbed. In pursuing this effort, professional pipemen, who had actually worked in the mine, were recruited to operate the equipment. This segment of highway, which took six years to complete, is now part of Highway 299 which connects the Sacramento Valley with the coast.

The La Grange mine opened again in 1940 in a kind of snipping process to work off the easily available gold without a major outlay of cash and was in limited operation by the beginning of 1941. From the first of the year until July 1 and later, during a short period from December 16 to the 31st, the yield from 113,000 cubic yards of gravel was 757 ounces of gold totaling $26,495 (at $35,00 per ounce). The following year, operating only from January to July, the mine produced about $20,000. It was then forced to close down with the outbreak of World War II.[60] In later years, sometime in 1951, a dragline dredging operation was established to rework the tailings of this once great hydraulic mine—reportedly resembling, somewhat, a vulture picking at the remains of a magnificent grizzly bear. This operation continued for a time with moderate success.

During its colorful history, the vast La Grange Hydraulic Mine—once the world's largest—removed over 100,000,000 cubic yards of gravel and produced an amazing $8,000,000 in gold bullion.[61]

The Spring Valley Hydraulic Mine

Less than two months after the discovery of gold by James Marshall at Coloma, another California pioneer, John Bidwell, also found color some distance away, in what would later become Butte County. Bidwell, a close associate of Captain John Sutter, learned of the events at the mill shortly after Marshall's arrival with his astounding news. Upon visiting the discovery site a short time later, Bidwell was impressed with the similarity of that gold-bearing region with the familiar banks of the Feather River near his Rancho Arroyo Chico, prompting him to hurry north. Arriving at the Middle Fork, Bidwell prospected with success at a place later to become the town of Hamilton (not to be confused with Hamilton City), however, within a month he would find even richer ground a short distance upriver at a site destined for fame as—Bidwell Bar. In this locality rich auriferous Tertiary gravel deposits were soon located and the first hydraulic mines were introduced in 1854. By 1870 over 40 mines were engaged in hydraulic operations in the near vicinity.

But the hydraulic mines that would gain world fame in Butte County were located about eight miles north of the town of Oroville on the celebrated Table Mountain, a noted California landmark. According to Rossiter W. Raymond, it is the finest example of a basaltic covered tableland in the state. It is actually a visible ancient riverbed covered with rich auriferous gravel. It comes into view on the north bank of the Feather River and extends along the eastern rim of the Sacramento Valley to the gold rush town of Cherokee.[62]

Although gold had been discovered in the region as early as 1849, the name Cherokee was given to the area by a group of Indians of that tribe who had migrated from Georgia in 1850 and arrived in Butte County three years later. Profitable gravel deposits were discovered by these early settlers paying from $20 to $50 per day per man. The early claims were limited to 100 square feet in accordance with the mining laws in the district, and were worked with rocker, long tom and sluice boxes. These claims were later enlarged to measure 150-feet long by 100-feet wide. A small mining camp constructed of canvas and whip-sawed lumber soon followed and by August 17, 1854, a post office, with the designation Cherokee, was established in a small business section with stores, saloons and a hotel. The first tunnel or drift mine did not get off to a very auspicious start. The Charity Blue Gravel Claim began operations in 1856, and after a heavy investment, had not reached the pay streak after 11 years of intensive work. Perhaps because of this failure and others like it, after 1858 all new mining in the area was exclusively worked by the hydraulic process. Because of the size of the claims and the scarcity of water, many of the owners were often negligent in obtaining their annual renewal as required by California law. As a means of protection from the possibility of forfeiture, many of the miners banded together under the leadership of Charles Waldeyer into a loose confederation known as the Butte Table Mountain Consolidated Mining Company. Whenever a discouraged miner moved on to more promising prospects, his claim was purchased by the company—some for as little as five dollars. The new organization soon controlled over 1,000 acres of mining ground, although some claims were not contiguous.

In a move to acquire development capital, the association, by a complicated legal maneuver, was dissolved and repurchased, forming a new corporation known as the Cherokee Flat Blue Gravel Company with Charles Waldeyer serving as superintendent.[63] It should be noted, however, that a remnant of the Butte Table Mountain Consolidated Mining Company remained intact.

By 1870 there were a number of major hydraulic claims in the close proximity of Cherokee, including: the Cherokee, Spring Valley, Eureka, Welsh, and the Cherokee Flat Blue Gravel companies. These mines confined their operations solely to the winter months as they were entirely dependent on rainfall for their water source. The isolated and elevated position of Table Mountain made the introduction of water a very difficult and expensive operation.[64]

The Cherokee Company, incorporated in 1867, owned rich holdings west of Sugar Loaf, between the Spring Valley Mine and the Blue Gravel Company. In an effort to maximize their water supply, the company invested $150,000 in the construction of ditches and reservoirs on Table Mountain. After building three miles of earthen embankments with a base of 60 feet and an average height of 20 feet, a basin was created covering 100 acres. The company also excavated a drainage tunnel which ran from Sawmill Canyon through the mountain and out into Sawmill Ravine.[65]

The most prominent mine at Cherokee Flat in 1870 was the Spring Valley Mining & Irrigation Company which had recently been incorporated by a group of San Francisco investors led by Egbert Judson, who was also associated with the Milton Mining & Water Company. This new company had plans to immediately bring water to Table Mountain.

Next was the well-financed Cherokee Flat Blue Gravel Company reported to have ownership of 1,000 acres of valuable gravel deposits with over $160,000 in permanent improvements. Both of these rival companies were in competition, seeking the rich, blue gravel layer under Table Mountain. Each was working on opposite sides of the hill—the former using two Fisher Knuckle Joint hydraulic monitors while the latter was driving a bedrock tunnel. The Spring Valley Company reached its goal first, the *Mining and Scientific Press* reported that in March of 1871 they had struck very rich, blue gravel in the "direction of the Blue Gravel Company's line running under the main Table Mountain." The same article noted that the Cherokee Company was making arrangements to bring the waters of Butte Creek to their mine by the end of summer.[66]

But the Spring Valley Company, in 1870, succeeded in being first to bring water, from an outside source, to the mines at Cherokee Flat. In this endeavor they were responsible for one of the greatest engineering achievements of the 19th Century. The project involved obtaining the water rights from an area high above a reservoir in Concow Valley which was formed by constructing a large earthen dam across the valley floor. From this point, the water flowed through a ditch to Yankee Hill, there it was conducted across the ravine of the west branch of the North Fork of the Feather River to the opposite mountain where it entered a canal leading to the mines at Cherokee Flat. Crossing this ravine required the construc-

THE SUMMIT　　　　185

Spring Valley Hydraulic Gold Company Mine, Cherokee, California.
From Wells & Chambers, *History of Butte County*, 1882.

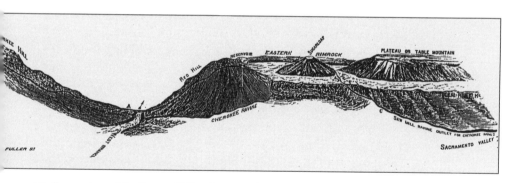

Bird's-eye view of the Spring Valley Company's works. A—Iron pipe of Spring Valley Company, diameter 30 inches, length 14,000 feet, vertical height on Yankee Hill, 900 feet; on Red Hill 830 feet. B—(to right and below Sugarloaf) tunnel of Spring Valley Company. C—tunnel of Cherokee Flat Blue Gravel Company.
From Rossiter W. Raymond, Fifth Annual Report, 1873.

tion of the greatest inverted siphon ever attempted at that time. The success of this pioneering effort paved the way for future, even larger, inverted siphons, such as the one constructed by the La Grange Company many years later.

The inlet of the pipe was 150 feet above the outlet, with a vertical height from the lowest point to grade line of 900 feet. The pipe was 30 inches in inside diameter and was designed to carry 1,900 miner's inches of water. Its thickness ranged from one-quarter of an inch to three-eighths of an inch. At the lowest point it achieved a pressure of 887 feet or 380 psi—the greatest pressure that water had ever been carried anywhere in the world. The water entered the upper part of the claim from a ditch into a pressure box with a sand box attached. It was conducted into the cistern by means of an elbow dipping below the water to prevent the possible entrance of air. Fifty feet from the inlet a stand-pipe was installed to allow the escape of any air that could possibly get into the pipe. At intervals along the surface of the inverted siphon, valves were also installed made with floats to allow for the escape of air. When the water entered the pipe for the first time great care was taken while it was being filled and the air valves were carefully watched, for no engineer had ever worked with this kind of pressure before.

As the pipe filled, the progress of the water could be traced as the air valves blew-off with a loud report, allowing the accumulated air to escape. When the column of water rose to the point of discharge into the ditch, not a single air bubble was detected. In the event the water would ever be shut off for any reason, causing a vacuum, these valves would also prevent the collapse of the pipe from atmospheric pressure by allowing air to be admitted into the pipe. The pipe was supported over the ravine by means of a trestle at a height of 70 feet. All of the rest of its length was buried in a trench covered with earth at a depth of five feet to prevent undue expansion and contraction. For the entire extent of the ditch, there were three sections of iron pipe with the inverted siphon segment achieving a length of 13,100 feet.

The 30-inch section of pipe, totaling nearly 14,000 feet, was shipped by barge from the Risdon Iron Works in San Francisco to Benicia, then hauled overland to Cherokee Flat. This accomplishment was no small feat in itself and was followed with keen interest by most of Butte County as its progress was closely chronicled by the local newspapers. Most of the pipe, however, a total of 87,000 feet, was manufactured at the site, the iron plate again furnished by the Risdon Iron Works. It was fabricated at the rate of 1,100 feet per day, giving employment to a large labor force using machinery designed expressly for the job and consuming as much as 30 tons of iron daily. The pipe was laid in lengths of 23 feet, which were riveted one to the other continuously. Manholes were constructed every 1,000 feet to allow for easy access for workmen. A steam riveting machine was employed for nearly all of the project which, according to reports, gave better results than hand labor.[67]

The engineer in charge of the design and construction of this monumental undertaking was well qualified for the task, having recently completed the waterworks for the City of San Francisco. Hermann Schussler, a graduate of the Prussian Military Academy at Oldenburg and trained in the engineering schools of Zurich and Carlsrube, had arrived in San Francisco in 1862 where he was hired by the Spring Valley Water Works to bring water to the city, first as an assistant and then chief engineer. After completing the job at Cherokee, Schussler would later gain additional fame by successfully bringing water from the Sierra to the mining towns of the Comstock.[68]

Many San Francisco capitalists had been steered away from investing in the Spring Valley water project by the gloomy forecasts of so-called authorities, who proclaimed an inverted siphon of this magnitude an engineering impossibility. Even respected mining engineers doubted its feasibility. As Charles Waldeyer later stated: ". . . men of good sense and some engineering capacity expressed fear that air would be carried down with the water to the lowest point of the pipe or siphon, and collect there finally to a great extent under the pressure of two immense columns of water, and might explode at any moment as in an over-charged air gun."[69] During a preliminary test, these doubts seemed justified when a leak was detected. But it was soon determined that the cause was merely a faulty manhole and not a design flaw, and was quickly repaired. On December 19th, all hands, with varying degrees of apprehension, along with most of the town, crowded to the site awaiting the first arrival of water. The Butte *Record* recalled the great event in its Christmas Eve edition.

> On Monday last, water was turned into the large iron pipe to plunge down a depth of 800 feet, and force its way up the same distance to its point of discharge. It was calculated that it would take four hours to fill the pipe so that it would commence to discharge, and it might take much less time to cause a break. How anxiously passed the time, as the lower air-valves, by the approaching water marked its rise by distances of 100 to 200 feet. Three hours have passed since the water was turned in. Confidence and anxiety are depicted on the faces of those in charge of the work—forty minutes pass —from its mouth pours a stream, half filling the cavity of the pipe. Anvils are fired, and impromptu rejoicing are kept up during Monday and Tuesday, when we visited Cherokee to see with our own eyes and hear with our own ears. A friend that accompanied us, declared that every man in Cherokee had his nose skinned. They can well afford to rejoice. For over twenty years the richest mining section in the Golden State has been dependent upon the winter rains for water for mining purposes. Now they have a living stream that shall cause the placers of that section to yield up their hidden treasures at the touch of its magic wand.[70]

Although overshadowed by Spring Valley's monumental operation, the Cherokee Hydraulic Mine was busy during most of 1871 in preparations for bringing water from Butte Creek. This project included another pipeline 34 inches in diameter and 3,780 feet long, incorporating an inverted siphon with a vertical drop of 680 feet. The greatest thickness of the piping was three-eighths of an inch at a point where the pressure equaled 295 psi. The length of the new system, including the headwaters of Big Butte Creek and the west branch of the Feather River, aggregated 52 miles, the average width was five feet and the depth, three and one-half feet. With the new water supply the mine ran two and sometimes three, seven-inch monitors in the summer and five in the winter.

The following year, the Spring Valley Company completed a 1,500-foot bedrock tunnel making it possible to work the claim to a depth of 500 feet. At that time the company employed about 80 men with three miles of sluices paved with stone and numerous undercurrents. A month after its completion it was reported that the mine had concluded a clean-up valued at $50,000 making a total production of $200,000 for the season

In February of 1873 the Cherokee Company and the Spring Valley Mine, whose works were adjacent to each other, completed a merger ending an intense and sometimes bitter rivalry. The former company owners had little choice because of their chronic need for more water than they could otherwise obtain. Through this joint effort, the new combination, after a total expenditure of about $750,000 in improvements, was anticipating production to run around $2,000 daily. It was capitalized at $4,000,000, valued at about 75 percent of the stock, and paying nearly two percent monthly. The officers of the company were: Egbert Judson, President; Mr. R. C. Pullium, Superintendent; Louis Glass, Secretary; Trustees, E. Judson, William Gregory, Isaac E. Davis, R.C. Pulliam and Richard Abbey. The financial statement, ending in July, showed assets of $4,344,490 and liabilities amounting to $143,869, it also disclosed that during that same fiscal year, the company had paid out $150,000 in dividends.

At that time it was determined that the ancient channel extended from about eight miles north of Cherokee Flat to Thomson Flat in the south and was from two to four-miles wide. According to the *Mining and Scientific Press* "... it can be justly claimed as one of the richest and most extensive gravel deposits in the state." It was also noted that the company was operating six hydraulic Chiefs equipped with six-inch nozzles, each capable of discharging 1,000 inches of water. The new company, which retained the name Spring Valley Mining & Canal Company, controlled about 1,500 acres of ground containing pay gravel to an average depth of 300 feet. On the line of their ditch they had over four miles of iron pipe, 30 inches in diameter. The mining property operated a staggering ten miles of sluices, varying from four to six feet in width.[71]

By the following year the consolidated companies completed a ditch, formerly owned by S. L. Dewey, 18 miles in length, extending from the west branch of the Feather River. This new addition served as a feeder for the Butte Creek Ditch—additional reservoirs were also constructed. By then the owners had expended in the works, flumes, reservoirs and water privileges over $1,000,000. For the year ending July 1874, the sum of $476,112 in gold was washed out and shipped. This amount included the largest gold bar ever made, valued at the mint at $71,273.15. When describing the historic event of the delivery of the record gold bar, the Butte *Record* reported:

> The boys made some little display over their trophy in bringing it down from their mine to Oroville. The vehicle containing it also conveyed several armed men with Henry rifles and revolvers and derringers. This was proceeded by another carriage occupied by Uncle Wm. Gregory [mine superintendent], and another gentleman, armed with a Henry rifle and a double barreled shotgun charged with buck-shot. In the rear was another armed guard. On arriving in town, the cavalcade halted at the banking house of Rideout & Smith to deposit their treasure. While in the bank and in the endeavor to remove the caps from the shotgun it was discharged, the contents carrying away a portion of the wall, and a map hanging thereon. Some think this was done on purpose, and in celebrating their successful clean-up. Whether so or not, it clearly demonstrated that, had Uncle William ever emptied his gun into the person of a footpad, who might imagine himself strong enough to walk off with the bar, the Coroner in holding an inquest would have struck an extensive lead mine. The bar was safely landed in San Francisco, and was on exhibition among the attractions of Woodward's Gardens on Sunday last. The Cherokee mines with their present facilities for working, will yield in the neighborhood of $50,000 per month, or $600,000 per annum. Scarcely a scar has been made upon the grounds of the Company.[72]

Unlike their counterparts in Nevada, Placer and Yuba counties, the mines at Cherokee, except in the very early 1850s, seemed to have shunned the hiring of Chinese labor. The reason for this, no doubt, was the intense anti-Chinese feeling in all of Butte County and particularly in the town of Cherokee. There the citizens formed a Caucasian League solely for the purpose of demonstrating their anti-Chinese feeling and also to warn mine owners away from the temptation of hiring non-white labor. Most community activities in Cherokee were held at Caucasian Hall. In any event, the town of Cherokee remained one of the few mining camps in Butte County without one Chinese inhabitant. Portuguese, on the other hand, appear to have dominated the labor force in the Cherokee mines. The only exceptions to the Chinese ban were on those occasions the mines subcontracted certain services to other companies.[73]

The Spring Valley Mine employed 160 men the year around with an annual payroll of $85,534. Total expenses for the year were $125,000 with an expenditure of $13,309 for quicksilver alone. The duty of a miner's inch at Cherokee was reported to be 5.5 cubic yards compared with the average for large hydraulic mines of not over three cubic yards. The ground being worked had a layer of blue gravel from 10 to 15 feet in depth, with the most valuable gravel yielding from one dollar to eight dollars per cubic yard. It was not cemented and could be removed by hydraulic streams. It was not unusual at Cherokee, when working banks 450 feet high, to cause caving that would bury a seven-inch monitor located 400 feet from the bank.[74]

Water supplying the mining property was brought in two ditches, 60 miles in length, from Butte Creek and the west branch of the Feather River. The ditch from the Concow reservoir was six feet wide at the bottom and eight feet wide at the top and four feet deep. The combined ditches ran a constant stream of 2,200 inches of water. A unique feature of the Cherokee ditches, unlike their contemporaries in the higher elevations, was the fact that they were patrolled by Kanakas from the Sandwich Islands (Hawaii). Although they were never considered good miners, they were great swimmers and had a fine aptitude for diving into the canals, swimming both above and below the surface, inspecting the banks for leaks or damage. Superintendent of Ditches, John Moore was highly pleased with their work. Another peculiar feature of the claim was the fact that a rather large amount of diamonds had turned up in the sluices, most of them small and of little value. One, however, weighed two and one-half carats and another was sold for $250.[75]

In the early winter of 1875 the Spring Valley Company, after a clean-up of just a portion of the head sluice, sent a gold bar to the mint at San Francisco valued at $40,000. At the height of the winter season, the company was then operating six to eight little giants. A few weeks later the *Record* reported, yet again, two more bars had been shipped, making a total for the month of something over $73,000. It was noted that the company had secured a monopoly on the waters of the west branch of the Feather River, in addition to their former rights to Butte Creek waters and those of the Concow Valley.[76] During that same year the Welsh Company Mine, which controlled a small but rich group of claims on the side of Sugar Loaf Crest, and the Table Mountain Consolidated hydraulic mine were both added to the Spring Valley holdings, forming a new corporation called the Spring Valley Mining & Irrigation Company with a capital stock of $5,000,000. The two principal officers were Richard Abbey, President, Louis Glass, Secretary and M. R. C. Pulliam served as Operational Superintendent. A majority of the stock was still held by Egbert Judson and a group of San Francisco capitalists.[77]

In 1876 when Rossitor W. Raymond visited Cherokee, he noted that the Spring Valley Mine employed 150 laborers at $3.00 per day, working on 250 to 300 acres of paying ground, 100 feet in depth, with three openings. He found three miles of triple lines of sluices—two were six feet wide and the other, four feet. They were constructed in such a manner to allow any portion to be shut off for a clean-up at any time. In addition there were 24 undercurrents ranging in width from six feet to thirty feet. At that time nine giants were operating under a head of 250 feet. Raymond found that the seven-inch nozzles could discharge 1,000 miner's inches; the six-inch nozzle, 700 inches; the five-inch nozzle, 500 inches.[78]

The natural course for the tailings washed from the Spring Valley and Cherokee mines was down the channel of Dry Creek, which bisected an area of small ranches of mostly fruit orchards. Periodically, during very wet seasons, the slickens from the mines would overflow the banks of Dry Creek and the orchards would be inundated by tons of mud. As early as 1872, A. J. Crum, a valley farmer, took strong exception to having his orchards covered with mining debris and brought suit against the Spring Valley Company. Like similar early lawsuits, which were to follow, the decision of the court held that since other mining companies were also dumping debris into the Dry Creek system, it was impossible to determine the degree of guilt of the Spring Valley Mine.

In an effort to minimize the damage to the farmlands and to quash any future lawsuits by the farmers, the company in 1875 adopted a policy of purchasing all the ranches through which the creek flowed, comprising 23,000 acres of agricultural land in the Rio Seco district. Before the end of the year the company had constructed a canal 400 feet in width which consisted of two channels 30 feet wide and 350 feet apart, with a fall of four feet to the mile extending a distance of 32 miles. The tailings flowed into two restraining dams 1,800 feet wide, covering 12,000 acres of the land located in the lower basin of Sawmill Ravine. The dam—more aptly called a levee, was nothing more than an earth fill six to eight feet in height lined with brush which was eventually filled in with tailings. The company spent $400,000 for the right-of-way and contracted to have 200 men employed in the excavation effort which was accomplished using "steam horse scrapers." The total cost of the project, including the ranches, amounted to $582,000.[79]

In an interview with mining engineer A. J. Bell, the Chicago *Mining Review* printed an update on conditions at Cherokee in 1880. " The Spring Valley Hydraulic Gold Company has 138 men employed in its mines at Cherokee, Butte County. The extent of the water supply is 2210 inches of water per diem. Six thousand cubic yards of dirt is the average day's work. Five banks of very rich ground are opened ready for piping. A greater portion of the ground worked, especially down towards bedrock, is blue gravel and rotten boulders. Tanks are being put in with a view of saving the black sand. There is an enormous quantity of this in the claim."[80]

The abundance of black sand at Cherokee led to the rumor that Thomas A. Edison had plans to establish a state-of-the-art reduction mill at Cherokee to work the black sands for platinum. The proposed company even had a name—the Edison Ore Milling Company, but it came to naught and was eventually dubbed a hoax.[81]

Early in 1881 the last and most important consolidation at Cherokee Flat took place with the union of the Spring Valley Company and the Cherokee Flat Blue Gravel Company. The official name was altered slightly to the Spring Valley Hydraulic Gold Company. The merger was brokered by Major Frank McLaughlin who organized a syndicate of New York capitalist that purchased both companies. The new organization was incorporated under the laws of the state of New York. Charles Waldeyer, long time superintendent of the Blue Gravel Company, was retained in the same position by the new corporation. At that time the company owned four reservoirs: Concow, Wasteweir, Hutchinson, and Round Valley holding 355,000,000 cubic yards of water. They operated 16, (some accounts say 10) hydraulic giants ranging in size from five-inch nozzles to seven-inch nozzles, constantly at work 24 hours a day, illuminated at night by electricity. To keep such an operation functioning, 40 million gallons of water were used daily with a payroll of 180 men.[82]

Later that year the new company, for the purpose of opening up new ground, proceeded to construct a bedrock tunnel 80 feet below its predecessor. The Butte *Journal* gave this accomplishment full coverage.

> [The tunnel] is 10' x 10' in the clear, and over three thousand feet long. The mouth of the tunnel is located at a point in Sawmill Ravine just below the crossing of the Miocene Ditch. The tunnel was driven from five different places. At a distance of 1804 feet from the mouth of the tunnel a shaft was sunk to a depth of one hundred and forty feet and the tunnel was driven in both directions! At another point, two thousand nine hundred and sixty-three feet from the mouth, another shaft was put down one hundred and four feet deep and work was also begun in both directions. The upper end of the tunnel is ninety-six feet below the surface of the worked ground in Sawmill Ravine, and forty-one feet in the bedrock—The gravel at this point being fifty-five feet deep, as determined by the shaft. . . . To accomplish this work two Ingersoll "Eclipse" drills — two in each header. These were furnished by Messrs. Reynolds and Rix of San Francisco.[83]

The greatest days at Cherokee appeared to have peaked out in the middle 1870s. According to Waldemar Lindgren, the most productive year was 1875 when the company shipped $406,900 in gold to the mint in San Francisco. From that time on production gradually declined reaching only $219,500 in 1878, (Wells and Chambers place this figure at $315,000). But the '70s as a whole had been a period of great productivity. In the Spring Valley claim alone, during the ten-year period from 1870 to 1880, the gold product was $2,650,809.[84]

By early summer 1883, the Spring Valley Company faced yet another lawsuit. Allen S. Noyes charged the mining company with dumping debris upon the plaintiff's land which was situated on the border of Butte Creek basin about 40 miles from the Cherokee mine. The defendants claimed, on the other hand, that no water reached the plaintiff's land other than what would naturally go there. They further maintained that the discharge from the mine after being strained of the sediment by brush dams in the canals and reservoirs, flowed pure into Butte Creek. The decision in this case was handed down the following October favoring the plaintiff with Judge Keyser, granting an injunction against the Spring Valley Hydraulic Gold Mining Company. This was a severe blow to the owners necessitating an expensive reworking of their restraining dams.[85]

After the Sawyer Decision in early 1884, like so many other hydraulic mines, the fortunes of the Spring Valley Mine took a sharp downturn. Although the company was already in possession of a restraining dam, new restrictive measures required expensive modifications, and production and profits continued to decline. By 1887 the Spring Valley Hydraulic Gold Company was in serious financial condition with many outstanding debts, including unpaid wages to their own employees. Fortunately, by April of that year, the company was able to obtain the services of another superintendent, the flamboyant and well-known William Gregory (Uncle William), who succeeded in reversing the downward trend. By November he had made three clean-ups: the first, for $45,000; the second, for $31,000; the third, for $50,000, for a total of $126,000. On December 10th, the *Mining and Scientific Press* was able to report that William Gregory had just paid off all the worker's back pay and added ". . . if they keep Gregory the mine will be out of debt in two years."[86] It appears the mining journal's prediction was based on sound reasoning for a year later it reported: "Superintendent Gregory shipped a $62,000 gold brick, the result of a three-months run and the mine is nearly out of debt locally and after a few more clean-ups will be on a 'solid basis.' The payroll is about $11,000 per month and is the best paying institution in Butte County."[87]

The Spring Valley Mine had been extremely productive. Prior to 1870 the mines that it had absorbed collectively had an estimated total production of $5,000,000. The value of bullion mined from 1870 until July 1886 totaled $5,008,208. During that same period expenditures ran as follows:

Reservoirs, ditches and pipes	$ 510,820
Mining plant and tunnels	199,780
Mining ground purchased	419,396
Land purchased	461,536
Cost of debris canal	270,811
Total	$1,862,244 [88] [$1,862,343]

But in 1888 work at the mine was completely halted because of litigation requiring the construction of a new restraining dam. It was opened again a year later, but on a very limited basis. According to an October 1889 article in the *Mining & Scientific Press*, William Gregory, ever since he was hired at the mine, had been urging that hydraulic operations be discontinued and mining proceed strictly by drifting. The article even related that a tunnel was to be run from the west side of the gold-bearing bank, adding, "prospects look good."[89] When the mine was reopened, however, it was still operated hydraulically but with a major difference— its massive water system had been sold for hydroelectric power to the Spring Valley Water Company of San Francisco, crippling its output immensely. Only natural drainage from the winter rains was then available which, during normal years, would keep operation going for a period of six months. Three, seven-inch nozzles were then in use, but only two could operate at one time. "Bank blasting" was instituted to assist in caving large masses of material for sluicing. Only about 12 to 16 men were then employed. The gravel was sluiced down a bedrock cut, followed by 100 feet of sluice boxes paved with boulders and containing mercury. From there it was washed down an inclined shaft 125 feet long to the main drainage tunnel cut through the rock for 3,400 feet. It then flowed down a ravine and into the impounding dam.[90]

In 1890, according to a report of the State Mineralogist, the Spring Valley Mine consisted of about 150 acres which had been worked

Hydraulic diggings at Cherokee Flat as they appear today. Author's Collection.

from the surface to bedrock for a depth of 500 feet. Another 150 acres still contained unworked ground to within 15 feet of bedrock. A portion of this latter area was under lease to a group of Portuguese miners. At that time the idea of drifting any part of the mine was not considered feasible. The employees at the mine were mostly Portuguese and were working at a wage of $2.50 for a ten-hour day.[91]

The once world famous Spring Valley or Cherokee hydraulic mine continued operations under these confined conditions for a number of years until finally completely shutting down. Estimates of total production have ranged from $10,000,000 to $15,000,000 with Lindgren's figure of $13,000,000, recorded in an official report in 1911, probably close, making this operation the most productive hydraulic mine in California history.[92]

The North Bloomfield Gravel Mining Company

The premier hydraulic mine of the San Juan Ridge, if not the nation, was the Malakoff owned by the North Bloomfield Gravel Mining Company of Nevada County. It could, perhaps, also be ranked as the quintessential California hydraulic mine because its history so closely paralleled that of the industry itself.

Mining in the area to become known as North Bloomfield, according to legend, was first recorded in 1852 when two Irishmen and a German discovered rich placers in the vicinity of the future Humbug Creek. Shortly after this event the trio found they were in need of provisions and one of the Irishmen volunteered for the trip to Nevada City. The following morning, after being dutifully admonished by his partners not to reveal the location of their strike, he proceeded to town. Upon completing his purchases he stopped by a saloon for a quick drink before continuing on his way. At this point a familiar, predictable and oft-told scenario unfolded. The Irishman proceeded to get drunk and inadvertently revealed the secret of the mine. The following morning, after sleeping off his binge, he was followed to camp by nearly a hundred eager miners who feverishly prospected the creeks in the area with little success. Finally, in disgust, the disgruntled horde returned to Nevada City, naming the location—Humbug.[93] That particular appellation, incidentally, was rather popular during the gold rush, Erwin Gudde lists nine Humbugs, with various suffixes stretching from Siskiyou to Tuolumne.[94]

With the advent of hydraulic mining, it was discovered that a rich Tertiary channel entered the area of Humbug from the direction of Woolsey's Flat then curved west through Lake City and Columbia Hill, and then South to Bridgeport Township. The ground in the vicinity of Humbug Creek was rapidly taken up by numbers of small hydraulic operators who proceeded to wash the friable layers of gravel with gratifying success. One of the mines, after encountering cement, even erected a 24-stamp, steam-driven mill. By 1855 a respectable mining camp, appropriately named Humbug, was established, boasting a canvas-roofed hotel grandly christened the Hotel de France by the proprietess Madam Auguste.

The following year, when a post office was established, the citizens of the town, out of civic pride, opted for the more refined name—Bloomfield. The post office department, however, took the liberty of revising the name to North Bloomfield to distinguish it from another town with the same designation in Sonoma County. The district continued to grow and by 1860 the town had 600 inhabitants, with a French population of no less than 130, engaged in mining and commerce.[95]

After 1860, however, North Bloomfield, like scores of hydraulic mining camps throughout the gold region, became a victim of the relapse in the industry, and mining closures became all too frequent events. By 1863, as we learned in an earlier chapter, the promising young town of North Bloomfield, located on the rich auriferous gravels of the San Juan Ridge, was nearly depopulated. But the district demonstrated a remarkable resilience and with the revival of 1865, it was well on its way toward recovery. That same year, an emigrant from Alsace, Emile Weiss, established the Humbug Brewery at North Bloomfield and another Frenchman, Marius Bremond, opened the French Hotel a short distance from town at a place called Malakoff.[96] Although there has been a good deal of speculation over the source of the name, "Malakoff," historian, Turrentine Jackson suggested that it had a French connotation and was inspired by the memory of an important engagement during the Crimean War where the French successfully stormed the great bastion of Malakoff at Sebastopol in the Crimea, 1854-1855. Apparently, the two regions bore some similarities.[97]

During the period of dormancy in the mines at North Bloomfield, one far-sighted Frenchman, Jules Poquillon, with the idea of attracting investors, began buying up the idle claims until, by the mid-'60s, he had acquired 1,535 acres of potentially valuable ground. Through one of his friends, Francois Louis Alfred Pioche, a San Francisco capitalist, who was associated with Samuel Butterworth, president and general manager of the famous New Almaden

Quicksilver Mining Company, Poquillon was able to persuade them both to form a group of investors and buy up the property.[98] It was soon learned that the list of other members of the company read like a "Who's Who" of San Francisco's financial giants, including capitalists William Ralston, L. L. Robinson, Thomas Bell, William Barren and R. Bayerque. The actual incorporation took place on August 8, 1866, when a certificate to that effect was filed in San Francisco. The property was described as being located in Humbug Canyon. Capital stock was issued in the amount of $400,000. Shortly after assuming his position, the new engineer, Eugene Verge, determined that the fall of the channel was so slight that it would be impossible to work the claims to any considerable depth. He proposed to run a deep cut through the gravel bed, and if necessary to run a tunnel at the lower end for the purpose of getting an outlet at Humbug Creek.[99] During the two-year period of development work, water was purchased from the Eureka Lake Company.

On September 28, 1866, the Nevada *Daily Transcript* ran one of its first stories on the new town: "The village of North Bloomfield, more generally called by residents 'Humbug' now wear quite a cheerful appearance. Considerable mining is being conducted by Butterfield & Company who have purchased extensive claims and are now running a cut and tunnel in order to open them." The newspaper's readers were then to learn, for the first time, of the French connection at North Bloomfield. "These claims are owned mostly by Frenchmen, and some forty are employed on the works. The hotel is filled with borders and French appears to be the ruling language. It may, with reason, be said that the French have taken Humbug."[100]

The top priority for the new company was to secure extensive water rights from the original ditch companies on both of the Yubas and their headwaters bordering the San Juan Ridge. Included in these transactions was the Rudyard Dam, owned by the Sierra Nevada Mining & Water Company, constructed in 1856-58. Because the controlling interest had belonged to a group of English investors, it was commonly referred to as the English Dam. The transfer was completed in 1867.[101]

During the summer months of 1868, the capital stock in the corporation was increased to $800,000 by issuing an additional 4,000 shares of stock valued at $100 each. This growth was necessary to fund the development of the water rights obtained the previous year—then considered the best in the county. Immediate plans called for the construction of a ditch capable of supplying 3,000 inches the year around. "The North Bloomfield Company, commonly called the French Company, are about to commence operations on their projected ditch," it was announced by a local paper. The route had been surveyed the entire distance with the stakes set to commence digging. It was determined to begin at the lower end, near North Bloomfield, and work up to a branch of the Yuba near Jackson's Ranch. The estimated cost was projected at $300,000. By August a contract had been let for the 23 miles of construction to J. R. Myers of San Francisco. This first section of the ditch included the difficult crossing over Humbug Canyon which required the installation of two iron pipes, 27 inches in diameter, and 1,200 feet long, spanning the chasm. The following November—17 miles of the ditch had been completed with only six miles remaining to arrive at the original destination. However, it was then decided to extend the ditch an additional six miles to Little Canyon Creek.[102]

Even before the ditch was completed, it had been determined that much more water would be required for the operations then proposed, and plans were already in progress to construct a dam at Big Canyon Creek located at an elevation of 5,450 feet. The first priority was to obtain additional water rights for Canyon Creek from just below the Jefferson Mill and then extending up the creek to a point above Bowman's Ranch. A notice to that effect was filed in the Recorder's office on March 8, 1869. That same month the first water from Little Canyon Creek and North Poorman's Creek arrived at the mine. By then the company had completed 40 miles of ditch which was supplying over 800 inches of water. The following month, the company completed the purchase of Bowman's Ranch, the well-known stage station on a branch of the Henness Pass Road. The previous winter, Superintendent F. W. Robinson had men post notices defining the location of the future reservoir on a number of trees about the site. Standing on the snow, they were nailed in position at eye-level. The following summer, much to their dismay, the signs were found to be 16 feet above the ground.

In less than a month following the purchase of Bowman's Ranch, the line of the ditch was being surveyed to Big Canyon Creek. The entire job was expected to be completed within 60 days. The completed ditch, to Little Canyon Creek, was running a full head of water which had been increased to 2,000 miner's inches, supplying enough water at the mine site to operate eleven pipes, where 40 men were then employed.

The Fourth of July 1869, was celebrated in grand style at the little settlement at Bowman's. The local people invited about 20 employees of the North Bloomfield Gravel Mining Company to the hotel to share in the festivities. A gala table was set with ample food and generous libations leading to numerous rounds of drinks toasting the success of the undertaking. It was a day to remember, unlike any other ever experience in that mountain fastness. In fact, Bowman's Ranch had never before seen such lively times. By August, the hotel was crowded with workers with the overflow living in nearby camps. The construction crew, which consisted of about 450 men, was under the able supervision of contract civil engineer J. W. Currier.[103]

By December, with the construction of the dam well underway, the hotel at Bowman's was deserted with the doors and shutters, and anything else worth the trouble, carried off. At the bottom of the dam, a cast iron discharge was installed, designed to let out 10,000 inches of water. One-quarter of a mile below the dam was a "catch water," which, paved with cobble stones, led the water to the ditch. During the dam's construction, 600,000 feet of lumber was utilized.[104]

That same month, mention of the dam project appeared in the *Daily Gazette*, in a rather disparaging manner.

> The dam now being built at Bowman's Ranch by the Bloomfield Gravel Mining Company will be 45 feet high and 240 feet in length. It will form a lake one and a quarter miles long by one mile in width in the widest place and when full will flow about 700 acres. The dam is made in the form of a V, the point extending in toward the reservoir or lake. The entire cost of the structure is estimated at $20,000. Heavy timbers, secured by iron bolts will be the principal material used in its construction. The dam will probably

last but a few years, as the timber being subject to the influences of the weather will soon decay.... Should the dam ever give way from the pressure of its waters would make havoc in their downward course and the town of Washington would be apt to suffer from the flood. It strikes us that it would be a matter of economy and security if it were built of stone and mortar instead of timber.[105]

Before year's end the reservoir at Bowman's contained a body of water one and one-quarter miles long and averaged three-quarters of a mile wide. The dam had been completed to a height of 45 feet. By then it was anticipated that later it would be raised to 62 feet. It was learned that a civil engineer by the name of Hamilton Smith, Jr. had been placed in charge of dam construction at both Bowman's and at the English dam site. It appears that the *Transcript* had a much higher opinion of Bowman Dam than the *Gazette*, describing it as built "in the most substantial manner. The ballasting alone required 2,000 cubic yards of stone."[106]

At the Malakoff Mine, toward the end of the year, the North Bloomfield Company ordered two of Craig's new, six-inch monitors. As the technology continued to improve, the old method of using a great number of the smaller, more primitive nozzles was giving way to the use of fewer large, high-volume monitors. The gravel was processed through a six-foot flume one-half mile long. Near the mine site at North Bloomfield three large reservoirs were also constructed each covering an area of from five to ten acres.

The line of ditch from Big Canyon to Little Canyon Creek was nine and one-half miles and was flumed for one-half of that distance. The flumes contained 1,200,000 feet of lumber. During the course of construction, great skill in engineering was required to cross two heavy bluffs and two river crossings at a height of 500 feet, each costing $10,000.

Bowman reservoir covered a beautiful mountain valley in Eureka Township. When the water began to fill the lake, the top of the flagstaff marking the site, was still visible and it was reported the old Bowman house could be seen floating about "in a lost and forlorn way."[107] Towards the end of the year, to keep abreast with costs, the company found it necessary to levy an assessment of five dollars per share of the capital stock. In a published legal notice, it was announced that any unpaid assessment would be delinquent by December 10th and a sale of unpaid stock would take place on December 28th.[108]

During the period of the construction of the Bowman Dam and the connecting ditches, the North Bloomfield Gravel Mining Company, under the direction of Hamilton Smith, had a propensity for hiring Chinese workers at wages ranging from $1.15 to $1.25 per day in preference to $3.00 per day, the going rate for white workers. This policy generated mounting animosity on the part of the non-Chinese miners who showed their displeasure by not only protesting verbally, but by repeated acts of willful destruction, principally accomplished by setting fires to the company's flumes. Such extreme demonstrations of costly vandalism were too much, even for the fervently anti-Chinese *Daily Gazette*, which protested the actions of the hotheads in an editorial asking for special consideration for the North Bloomfield Company. "The company have spent a large amount of money — some say $800,000 in this county, and with scarcely any return thus far."[109]

The antipathy towards the Chinese was hardly confined to the gold-mining districts of California, but was merely an extension of the anti-Chinese feeling that had existed statewide from the days of the gold rush. With the completion of the transcontinental railroad in 1869 and the recession of 1872, and its mass unemployment, the sentiment against the Chinese continued to swell, culminating in the Exclusion Bill passed by Congress in 1881 and implemented the following year.

In 1870, the North Bloomfield Gravel Mining Company obtained a majority of the stock in the Union Gravel Mining Company and a controlling interest in the Milton Mining & Water Company. Both of these companies were in need of additional water to work their rich mining ground. They had nowhere to turn but to the North Bloomfield Company which had obtained all of the available water rights serving the San Juan Ridge, not then controlled by the Eureka Lake & Yuba Canal Company. In return for their stock the Union Company obtained free water privileges while the Milton Company came into possession of the Rudyard Dam, valued on the North Bloomfield Company's books at $67,544. Also, an agreement by the company to construct a 44-mile canal from Milton, the location of the intake ditch, to North Columbia. The Milton Company completed the balance of the ditch which served their mines at Badger Hill, Manzanita Hill and Birchville, before terminating at French Corral. The Milton mines included some of the most valuable properties on the San Juan Ridge—the French Corral Mine, Kate Hayes and Troy mines, the Bed-Rock Mine, the Manzanita Mine, and the Badger Hill Mine.

On September 10, of the same year, the ditch from Bowman was finally completed with a capacity of 3,000 inches delivered to Malakoff and an additional 1,000 inches through a seven-mile branch ditch to Columbia Hill.[110]

The following month the *Mining and Scientific Press* ran a story on conditions at North Bloomfield, remarking that at nearby Malakoff, one mile from North Bloomfield, was situated the office and principal works, of the North Bloomfield Gravel Mining Company with a capital stock of $800,000, held by citizens of San Francisco. Among the officers of the corporation, it named A. M. Dobbins as general superintendent (the following year, the contract, civil engineer, Hamilton Smith, Jr., would be hired for this position). It then continued, "[They] own about 1,600 acres of valuable mining ground at North Bloomfield and have constructed a ditch from Big Canyon Creek a distance of 40 miles. The ditch cost about $500,000. Also, they have constructed a dam across Big Canyon Creek at Bowman's Ranch 65 feet high. They also own the Rudycan [Rudyard] reservoir, formerly owned by an English Company."[111]

In a later follow-up article, it was noted that by the fall of 1870 the Malakoff owned about $40,000 worth of equipment at the mine site including a network of sluices one and one-quarter miles long varying in width from 40 to 72 inches. When at full capacity the mine worked three monitors, each capable of carrying 1,200 inches of water under a head of 300 feet. The sluices were so arranged that the clean-up could proceed without stopping work. The main sluices were paved with stone, while the branches had riffles of wooden blocks. The grade of the sluiceway was six inches for each twelve-foot box.[112]

The *Daily Transcript*, several months later, informed its readers of the booming growth in the town of North Bloomfield: the Atlantic Pacific States Telegraph Company had announced plans to extend their lines into town; an Odd Fellows Hall was being organized; a new boarding house had been opened and lots were rising in value daily; the paper then conceded, "North Bloomfield will likely prove no 'Humbug' after all."[113]

Late in 1870, a series of four prospecting or exploratory shafts were started at the Malakoff for the purpose of finding the exact location and direction of the channel, the depth to bedrock and the value of the ground. With this information it could then be determined how best to develop the property. At shaft No. 1, bedrock was struck at 207 feet, 135 feet of which was blue gravel averaging 41 cents per cubic yard. At bedrock level, drifts were driven for 1,200 feet on the course of the channel, estimated to be 500 feet wide.[114] In June of the following year another rich strike was reported at exploratory shaft No. 3. It showed a continuous channel from shaft No. 1 to shaft No. 3. Commenting on these developments, a local newspaper observed: "The company have expended an immense amount of money, and they have now succeeded in proving that the amount was well expended and they have in their claims a full reward for all their enterprise and labor."[115] From the bottom of the four shafts, a total of some 2,000 feet of drifts were run and extensive locations made from North Bloomfield to Relief Hill. In the course of these excavations, 21,640 tons of gravel was extracted with a yield of $36,600 in gold. The total cost for the exploratory shafts and drifts was determined to be $63,965.[116]

At shaft No. 1, located in Malakoff Ravine which emptied into Virgin Valley Creek, a tributary to Humbug Creek, a substantial building was erected over the excavation. Inside was located the hoisting works, which were first run by a 40 horsepower steam engine, but later replaced with a 17-foot hurdy-gurdy wheel under a pressure of 200 feet. The building also housed a blacksmith shop. A three-quarter-inch wire rope was used for hoisting the tubs up the shaft, each weighing about 1,000 pounds. The dirt was then dumped into a large wooden tank and washed by hydraulic power through a 125-foot sluice, which was cleaned up every 24 hours. About 25 men were employed at the site.[117]

Meanwhile, during the month of October 1871, men were employed at Bowman Lake to burn the driftwood that had accumulated behind the dam. While engaged in this task, before the horrified gaze of the helpless workmen, a large flaming raft broke loose from the shore and drifted against the wooden dam burning it to the waterline. As a stopgap measure, temporary repairs were immediately made, but during a heavy flood in December, the restorations were swept away. Following this reversal, it was decided the dam should be permanently rebuilt and the height increased to 73 feet.

Working through the winter months, at nearly 6,000 feet elevation in the Sierra, could be extremely difficult. In January, Bowman Dam was inaccessible except on snowshoes, nevertheless, there were, at that very time, 25 men in the employ of the North Bloomfield Company working in the vicinity of the dam, but little could be accomplished. The snowpack was so heavy that year that operations were delayed until the following June before serious work was begun. By mid-month, 40 men were observed passing through Graniteville to reach Bowman Dam, where by then 45 whites and 150 Chinese were employed cutting timbers, blasting rock for ballast, tearing down the remains of the old dam, laying tracks, and making all necessary preparations to rush the work forward. But more time was needed to begin actual construction on the dam, because the water level was still too high for permanent work to begin. There were other problems threatening the project as well. According to a report in a local newspaper serious trouble had been brewing in the area: "The difficulty between this company [North Bloomfield] and the Eureka Lake Company a few miles above Bowman's still continues, the gangs of men of the two companies being at work close together building works to appropriate the waters of the lakes." It was later learned that previously, the two sides had been actually armed and were threatening each other. The trouble stemmed from the fact that the Eureka Lake Company, for years, had owned extensive holdings in the area and considered it their private reserve, and resented the North Bloomfield Company's incursions in, what they considered, their domain. Fortunately, physical violence was avoided and the two companies appeared to have settled their differences. When finally completed, the cost for the dam and ditches was determined to be no less than $750,000.[118]

The data obtained from sinking the four exploratory shafts at North Bloomfield resulted in a bold course of action that would lead to an unprecedented engineering achievement. In the spring of 1872, Hamilton Smith, Jr., the new superintendent and engineer, announced plans for a monumental bedrock tunnel nearly 8,000 feet in length that would be pushed to completion. On April 19, the company ran an ad in the Grass Valley and Nevada City newspapers, requesting the employment of a large number of hard-rock miners. Under the heading "MEN WANTED," it was phrased as follows: "First-class Hard Rock Tunnel Miners accustomed to the use of Single-handed Drills and Giant Powder [a trade name for dynamite]—steady work for several years will be given to good men. Apply with recommendations from old employers to: North Bloomfield Gravel Mining Company."[119] It will be recalled that labor relations with the company were already on a rocky basis over the Chinese question and the advertisement in the *Transcript* only exacerbated the situation. In the first place, the miners were not happy about working with giant powder, and actually contemplated calling a miner's meeting to protest the question. Less than four years had elapsed since the hard-rock miners at Nevada City and Grass Valley had openly rebelled against the use of dynamite, protesting its noxious fumes.[120] Most hard-rock miners in the Grass Valley-Nevada City area at that time were Cornish and known to be very fixed in their ways and detested "single-handed" drills, preferring their time-honored "double-jack" teams, reasoning it was safer to work in pairs in the depths of a mine. But what they appeared to take the greatest exception to was the requirement to produce recommendations from former employers—they claimed this demand was insulting. More likely these feelings were linked to their past conduct during the protest against the use of giant powder that resulted in a bitter but fruitless strike at the Banner,

Empire and North Star mines. Nevertheless, an average of 160 men were employed on the tunnel for over two years.[121] But even a year later, Hamilton Smith complained that he could not obtain the services of as many men as he needed to work in the bedrock tunnel. During this period the company was paying out from $40,000 to $60,000 per month for labor, machinery and tools.[122]

The overall plan of excavation called for eight double-compartment shafts to be sunk along the line of the tunnel—one for hoisting and the other for pumping. The ground proved very wet and nearly 500,000 gallons of water were raised every 24 hours. The pumps were operated by huge hurdy-gurdy wheels 18 to 21 feet in diameter, the waterpower furnished from the company's reservoir through 10,000 feet of iron pipe—the material alone costing $20,000. The shafts were four and one-half by nine feet in size, heavily timbered and sunk between 800 and 900 feet apart, with an average depth of 197 feet.

The first 6,000 feet driven from the mouth of the tunnel, were six and one-half feet high and six feet wide, the balance was eight feet wide by eight feet high. All of the work in the tunnel was done on a contract basis with 150 to 175 men making from $3.50 to $6.00 per day. Each of the 15 faces were treated as separate contracts, with the miners supplying the explosives and paying for lights. The company paid for tool sharpening and the removal of the muck. Each heading employed three shifts of two men each per day. A diamond drill was used during the last year of excavation—all the other 15 faces were driven by hand drills and giant powder. The work progressed, on average, over 100 feet per week. When completed, the fall from the mouth of the tunnel to its outlet at the South Yuba River was 500 feet, giving ample room for all of the dumps and undercurrents that were required. The tunnel was not driven in a straight line, but was engineered with slight angles from one shaft to the other. Upon completion, a survey revealed how extremely accurate the tolerances were maintained: in alignment, nine-sixteenths of an inch; in levels, five-sixteenths of an inch; in distance, one and one-quarter inches.[123]

At the time of the early construction of the tunnel the *Transcript* wrote an account of the changing conditions at North Bloomfield.

North Bloomfield is now a lively mining camp. The company is employing a large number of men, and on Wednesday the first of the eight or nine hoisting works on the line of their tunnel was started up at shaft No. 8. A new hotel has been recently opened for the accommodation of the men employed by the company. The hoisting works will be run by hurdy gurdy wheels, and in the eight shafts, when opened, they will work the tunnels from sixteen faces. When their ground, amounting to nearly 1,500 acres and running some 2,000 feet upon a channel of great width from which they took out over $30,000 in prospecting by drifting, is opened, the company will have one of the richest gravel mines in the country.[124]

In describing the precision and skill demonstrated by Hamilton Smith in his calculations for the tunnel, W. W. Kallenberger recalled: ". . . where the shortest of base lines at the collars of each shaft had to be so precise that large plumb bobs suspended on wires over 200 feet in length were placed in cans of water at the bottom of the shafts so that a minimum of oscillation was to be had."[125]

It is unfortunate that this same skill was not demonstrated by the management in its labor relations—or perhaps matters just got out of hand. On August 27, 1872, yet another incident was reported. Two new hoisting works, Nos. 3 and 4, were both completely destroyed by fire at a loss of $6,000. The fires were set simultaneously at locations 1,000 feet apart and represented a major setback for the company, particularly in loss of precious time.

The labor problems at North Bloomfield would continue until 1881 when the company finally announced a new policy of hiring only white workers (the reason for this change of policy will be dealt with in Chapter 6). In July, of that year, the *Transcript* could report: "The only obstacle in the way so far to complete harmony of action between the people of the mountains and the hydraulic mining companies has been the fact that some of the latter employed in certain departments Chinese laborers to the exclusion of white men who were willing to do an honest day's work for fair wages. But at last, thank God, an order has gone forth that no more Celestials need apply at either the North Bloomfield or the Milton mines."[126]

The tunnel was finally completed in November 1874, at a cost of $498,800. Total expenditures, by that time, were $1,979,760, of this amount $1,031,000 was spent for ditches, reservoirs and water rights.[127]

Beginning in 1875 and continuing into the following year, the dam at Bowman was increased in height to 96-1/4 feet. The work was accomplished by filling in a stone embankment on the lower side of the old structure faced with dry rubble stones. The downstream face-wall was 15 to 16 feet at the base and six to eight feet at the top. Three wrought iron pipes, 16 inches in diameter passed through the water-face of the dam. A separate gate was placed at the lower end of each pipe. They discharged into a covered timber flume seven and one-half feet wide and one and three-quarters feet high which passed to the lower edge of the dam and discharged with enormous force, falling thirty feet onto solid bedrock in the bed of the old stream where it flowed for several hundred yards before entering the ditch.

Improvements on the reservoir continued until 1879, making it possible to back up 907,000,000 cubic feet of water or 410,000 24-hour inches which formed a lake covering 530 acres. The dam was described as "extending from the left bank of a ravine to a bold island of concrete some 500 feet in height. Between this island or peak and the right bank of the ravine is the waste weir, built of huge cedar logs bolted together and filled in with rock. When the reservoir is once full, there is in good seasons a wastage here sufficient to supply quite a city. This season it is calculated that the unused water flowing over at this one waste weir is equal to more than 400,000 miner's inches."[128] The total cost of the dam when completed was $132,000.[129]

In his design of Bowman Dam, Hamilton Smith, Jr. demonstrated a genius for small but important details that would prove invaluable in years to come. "The North Bloomfield possesses in this reservoir an advantage over all other companies on the great ridge" the *Transcript* related. "When the snow falls in the ditch of this company, water is let on from the bottom of the great reservoir, and thus at once melts the snow—a feat which no other company can accomplish by the use of the cold surface water."[130]

The modern Bowman Dam with an obsolete metal flume in the foreground. Author's collection.

Finally by 1879, even more water was added to the system by means of a ditch and pipe which conducted the flow from Texas Creek to the main canal four miles below Bowman reservoir. The pipe was an inverted siphon 4,400 feet long, 17 inches in diameter made of riveted iron plate. The inlet was 310 feet above the outlet which sustained a pressure of 770 feet or 334 psi discharging about 1,250 miner's inches. Hamilton Smith now considered the system complete with a water supply sufficient for constant use for the entire year, except in very unusually dry seasons.[131]

With the completion of the tunnel and all of the development work finished at Malakoff, quite a change came over the town of North Bloomfield as described in a story in the *Transcript* and copied several days later by the *Mining and Scientific Press*.

> The town is not as lively as it once was. When the tunnel was being driven a large force of men were employed. Now the mine is run by comparatively few making a big difference in the prosperity of the town. Still the

residents seem to be prospering as well as in other parts of the county. We found everything running smooth at the North Bloomfield mine. There has been a hole washed to get down to bedrock, where the shaft from the tunnel taps the gravel, which is now about a quarter of a mile square, and the center has not yet reached bedrock. This space has been washed out during the past six months, which shows that extensive work is being done. The gravel pays well from the surface down as far as it has been washed and increases in value as the bottom of the channel is reached. The mine is now in splendid working order, and will from this time forth return to the owners liberal dividends for their great outlay. The ditch owned by the company supplies all the water used at the mine. There are four pipes being run now, and about 3,000 inches of water used.[132]

In less than a month following this article, the serenity of the town of North Bloomfield was shattered by the clang of bells, the shrieks of women, and the shouts of men—fire had broken out in the French Hotel and quickly spread to Skidmore's Saloon. But despite the efforts of a hastily organize bucket brigade, the flames were not controlled until they had also consumed the office of Doctor Farley and the adjacent barber shop.[133]

But the prosperity of the mine was unaffected by the tragic event in town, and after a run of just 22 days a gold bar was produced worth $15,800. With expenses of $7,800, this left a very satisfying net profit of $8,000—a harbinger of good times ahead. By the following year's end, the North Bloomfield Gravel Mining Company was able to declare its first dividend of one dollar per share after achieving a gross profit of $291,125. It should be pointed out that this gratuitous event followed 43 assessments totaling $1,545,000. The first 26 assessments had been levied on 40,000 shares at $19.50 which equaled $780,000, followed by 17 more levied on 45,000 shares for a total of $765,000.[134]

As a matter of interest, William Ralston, one of the original investors in the enterprise and, nationally renowned San Francisco banker, who tragically died in late August 1875, received no returns from his interests in the company. An item in a local newspaper speculated that his investment was $700,000 or $800,000, but his biographer, David Lavender, places the amount at the more modest figure of $250,000.[135]

The death of William Ralston by drowning, (generally believed a suicide) during an accustomed swim in San Francisco Bay, which happened immediately following the collapse of the Bank of California and the subsequent closing of that esteemed institution had interesting implications for the North Bloomfield Company. These events, also shed more light on the very close relationship between the North Bloomfield Company in the exercise of its controlling interest over the affairs of its fledgling associate, the Milton Mining & Water Company. The latter company owed a considerable debt, dating from the original purchase of most of its holdings in 1871, amounting to an annual payment of about $50,000. Both the North Bloomfield Company and the Milton Company had close ties and a substantial line of credit with the Bank of California which was greatly facilitated through their close relationship with Ralston. The Milton Company, especially, depended on it for necessary funds to meet their annual note.

In an inter-company (Milton) memo dated August 31, 1875, Hamilton Smith noted the difficulty he was having finding funds to pay the note after the suspension of business at the Bank of California. He expressed the opinion that he would have to seek help through a loan from some individual (possibly Egbert Judson). On a happier note it was his belief that the bank would be opened for business within ten days. He was also sure that D. O. Mills would take over its management. He then made the interesting comment that Ralston's investment in the North Bloomfield Company would be transferred to the Bank of California, and suggested that Mills would be very desirous of maintaining the good health of the mining company to protect the bank's own interests.[136]

It was the practice at the Malakoff to cleanup the upper part of the mine's sluices, which generally contained about 80 percent of the saved gold, every two or three weeks. But a clean-up of the entire system, which included the bedrock in the tunnel and all of the undercurrents, strung along the mountainside from the mouth of the tunnel to the South Yuba, was a major undertaking and was only done annually. The complete process required a week's time and

Undercurrents at the Malakoff Mine.
Courtesy, Searls Historic Library Collection.

involved shutting off the water supply at the head of the ditch, 40 miles away. The blocks of wood used for sluice pavement were specially prepared for the company by the sawmill and cut 13 inches thick. They generally lasted for four runs of two weeks each. Unlike many other hydraulic mines, the Malakoff did not burn the used blocks to recover the gold from the ashes. Some superintendents claimed by doing this they recovered enough gold to pay for new blocks. [137]

Boulder and pipe-clay blasting was a periodic activity at the Malakoff mine that was unique in its scope and preserved for history in this unusual eyewitness account.

> At the North Bloomfield mine I saw 40 men sent out immediately after the stream was turned to an unwashed area. The men commenced drilling, some single and some double, ... In the pipe clay masses of holes were drilled with augers. In a short time 180 holes were drilled and men congregated at the blacksmith shop in a body. Here the powder-men met them and served out cartridges and fuse equivalent to the number of holes drilled by each. With these the men quickly charged the drill holes and returned to the shop,

where 40 iron rods were heating. . . . At a second signal each man [with a red hot rod] fired his fuses and ran to a position of safety. 180 explosions quickly followed which sounded like the discharge of a part of artillery and the air was filled with flying fragments. . . . The blasting generally takes place at noon, after which the men go to dinner and all the streams are turned on to wash away the debris. When the men return again a new set of bowlders are exposed and drilled and blasting repeated.[138]

Following the annual clean-up of October 10, 1878, which completed the season beginning January 12th of that year, the company issued a report covering the period from 1870 to the close of 1878. In the four-year period from 1870 to 1874— 710,987 inches of water were used with a gross product of $96,700 which represented a yield of 18.8 per miner's inch. Skipping to the year, '76-'77—595,000 inches were used to produce $290,775—a yield of 48.9. The following year, which was the latest, although the bullion produced was the greatest ever at $311,277, more water was required to obtain it— 796,449 miner's inches, resulting in a yield of 39.1. The diminution was explained by the fact that the gravel channel worked that year was, comparatively, not as rich. During the same year, the company's cost for water was one-third less than in previous years—2.5 cents per miner's inch.[139]

Reviewing these same years from a different perspective may be more revealing. For example, according to mining engineer John Hays Hammond, from 1870 to 1874 the old washings of surface ground near Malakoff were estimated at about 3,250,000 cubic yards and yielded 2.9 cents per cubic yard, barely above the break-even point. From November 29, 1876 to October 13, 1877— 1,591,730 cubic yards were washed with a yield of 3.8 cents. However, during that same period the company washed 702,200 cubic yards of bottom gravel which yielded 32.9 cents per cubic yard—a very healthy profit. Bottom gravel extended from bedrock to a height of 65 feet while the depth of top gravel varied from a few feet to as much as 200 feet.[140]

By 1879 the bedrock tunnel had been extended, by means of a branch, an additional 1,006 feet, requiring the excavation of a ninth shaft. The addition was necessary to drain a portion of ground that otherwise could not be effectively worked. Upon completion, the extension lengthened the tunnel to a total of 8,875 feet. The new section, which was nine feet square, was driven by means of Burleigh air drills, requiring nine months to complete. The project averaged 112 feet per month, using only one shift of drillers and was reckoned at a cost of $24 per foot. Previous to the use of these drills, in selected sections where only hand power was used, three shifts of miners in the same tunnel only averaged 25 feet per month, at a cost of $27 per foot.[141]

In the following year, the North Bloomfield Gravel Mining Company arrived at what proved to be the pinnacle of its success before being materially hampered by the anti-debris movement. By 1880 the company was at the height of its productive capabilities, working over 100 men and using nearly 4,000 miner's inches of water. The Malakoff was capable of mining away about 50,000 tons of gravel per day. Excavations from 500 to 600 feet in width extended for 5,000 feet reaching a depth of 500 feet. At midyear the management announced another dividend, their sixteenth, in the amount of one dollar per share totaling $45,000.[142] Perhaps to cover these dividends or to augment their cash flow, the company appeared to have shortened their runs as noted by a news item in the *Transcript*. "Superintendent Perkins accompanied by a guard of armed men came over day before yesterday with another lot of bullion from the North Bloomfield hydraulic mine. It is the third shipment made by the company during the present month, and, judging from the efforts of the parties who carried the treasure boxes from the wagon into the express office on the several occasions, the three lots aggregate between $25,000 and $30,000 in value. As shown by the annual report this mine produced over $300,000 last year."[143]

We are fortunate to have preserved for us a contemporary eyewitness account of the Malakoff from the pages of the San Francisco *Bulletin* written in July 1879.

> We stand on the brink of the mine and try to fix the salient points in thought and memory before we descend into the great amphitheater, vaster

in its circle than the stony base of the coliseum. Around us are naked rocks and well scraped furrows, piles of pine wood blocks for use in the flumes, rusting joints of condemned water pipe and shops where soot-covered men are riveting joints of new pipe, sharpening drills at glowing forges, or in a thousand ways giving examples of the uses of iron. Higher than the red cliff of the mine are shaggy limits of forest which cover the northward ridges here consisting of an overlaying lava deposit extending toward Lake City. . . Far down under the highest cliff on the sloping bedrock, and half hid in shadow, are a multitude of men. The water has done its work here, and washed out all the loose earth and smaller rocks. Now the thing is to get rid of the large boulders, of ten weighing tons. They must be blasted into fragments so small that when the water is turned on here again they will be swept down and out through the tunnel. . . . The black pipe, three feet in diameter leads down the cliff and across the mine it becomes smaller and ends in a jointed elbow-like pipe with a moveable nozzle By laying the weight of a hand on the lever, this rim-like nozzle enters the edge of the stream and the weight of the water turns the machine to any angle desired. . . . It may be dropped till it foams at the operator's feet; it may be raised almost upright, or made to sweep the circumference of a circle. It is not hard work to manage one, but it requires much experience and judgment to know how to use the stream with greatest safety. The actual work of tearing down the cliff is hard to see, for there is a cloud of red foam hanging over the spot. You hear little rattling and slipping noises through the incessant roar, and a stream which seems ten times greater than could come out of the pipe, flows down the dripping pile, and so into the rock channels [bedrock cuts or ground sluices] which lead to the tunnel. . . . But the most interesting portion of the mine is at the outcome of the main tunnel which extends under Humbug Creek, two hundred feet down, and was worked from eight shafts. . . . The stream of water is so powerful that no man could stand up against it a moment. A Chinaman slipped in once when hardly half the stream was running, but the body was never found. . . . The water after leaving the tunnel is carried half mile or so in a flume so as to allow a chance for under-currents to collect more of the gold. These under-currents are side-floors or platforms of tight boards sloping from the center like the ridge of a barn [divided into two sections], at right angles with the main flume. The bottom of the main flume has iron bars fastened across it, about an inch apart. The rocks and most of the water goes over, but the heavier particles are carried out with sufficient water to fill a flume extending along the ridge in which the water flows in two wide shallow streams. . . . So powerful is the stream that the iron bars of which we have spoken, although four inches wide and one inch thick, are worn fairly in two, and bent and broken in a few weeks, by the boulders continually rolling across. .

THE SUMMIT 219

The great pit of the Malakoff Mine. Courtesy, Searls Historic Library Collection.

Malakoff Diggings Hydraulic Mine, North Bloomfield.

.. It is within bounds to say that this company has spent in the neighborhood of $3,000,000 before their mine began to pay. The larger part of this went for wages at three dollars and upward per day They are now paying from two and a half dollars for white labor. Part of their force are Chinese [$1.15 per day]. A working mine is the life, the very breath and substance of the adjacent town.[144]

Shortly after the close of 1879, the company released a statement of profits from the time of its original organization until the end of the decade.[145]

Year	Yield	Profit
1866-74	$ 218,073	2,233
74-75	83,079	20,072
75-76	200,367	98,476
76-77	291,125	148,172
77-78	311,277	140,696
1879	331,760	183,855
Total	$1,435,681	$595,444 [$593,504]

In 1880 the North Bloomfield Gravel Mining Company and the Milton Mining & Water Company, by an action of the joint board of directors, allowed the Milton Company to be absorbed in both ownership and direction by the older company. Corporate headquarters remained at the parent company's old offices located in San Francisco's financial district at 320 Sansome Street. Shortly, the two operations were employing 400 men at $2.50 per day (10 hours), on a 24-hour schedule. Together they produced about $1,000,000 a year in gold.[146] This merger could have only been possible as a result of the interlocking relationships of the two boards of directors. The Milton Company owned, perhaps, the most valuable mining property on the San Juan Ridge. For an idea of its productive capabilities, during the first nine months of 1878, it cleaned up $540,000–the best of any hydraulic mining company on the coast.[147]

The census rolls of the town of North Bloomfield serve as a good barometer of the health and wealth of its principal employer, the North Bloomfield Gravel Mining Company, which was then operating with a full crew of 100 men. They showed a slow but steady gain in population from 636 in 1870, to 1017 by the beginning of the next decade. Despite the problems the mine owners were encountering with the ceaseless lawsuits from the farmers in the valley, the town continued to improve. By the beginning of 1881, numerous new buildings were making an appearance in the business section of town, with Skidmore's saloon dominating the scene. It was a rather imposing two-story frame structure, with the lower floor devoted to an impressive barroom while the upper story served as a public hall. A new school house had also been completed the previous November and was well-located on a hill separating the town from the Malakoff mine. The community building, Cummings' Hall, with a spacious room measuring 60 by 28 feet, had become a popular place for dances and other public gatherings.[148]

But despite this outward display of prosperity, the debris problem was beginning to have a telling effect on the North Bloomfield Company. Later that same year it was forced to issue a new assessment on the company's stockholders—the first in a long time, and the board of directors also felt compelled to cease further payment of dividends in the then foreseeable future.[149]

In an apparent display of affluence, and to refute the claims of the anti-debris forces who were alleging that the hydraulic mines were all in serious financial straits, the North Bloomfield Gravel Mining Company, amid great fanfare, announced a gigantic clean-up which produced the largest gold bar ever cast in the United States. The product of just 20 days washing, the golden brick measured 17 inches long, seven inches wide and eight inches thick, weighing 511-1/2 pounds troy. With a value computed at $18.60 per ounce it totaled $114,250. Upon its arrival at Nevada City, it required four men to carry the huge bar. It was majestically perched on a light iron bed equipped with four handles, to enable them to ceremoniously remove it from the Wells Fargo stage. Before an admiring crowd of onlookers, it was then placed aboard the narrow-gauge to begin a well-publicized trip to San Francisco, via Colfax. Because of its great weight, there was little fear of the possibility of a theft while in transit. Upon its arrival in the city it was prominently displayed at the California Bank before being finally sent to the mint.[150]

As early as 1879 the Anti-Debris Association had taken direct aim at the hydraulic mines of the Yuba River by filing a complaint in the Yuba District Court, titled *The City of Marysville vs. The North Bloomfield Gravel Mining Company et al*. Although the suit was later struck down by the State Supreme Court and a compromise of sorts was put together by the state legislature with the passage of the so-called Drainage Act, the suit was again renewed in 1881. The following June, Judge Keyser made a momentous decision ruling that all companies named in the lawsuit immediately cease dumping debris in the Yuba River. On June 24, the Nevada City *Transcript*, woefully reported that the North Bloomfield, the Milton and Blue Tent mines were all at a standstill on account of Judge Keyser's injunction. The article noted that those mines produced about $3,000 per day in gold bullion when in operation and were wasting another $1,000 daily from the loss of water.[151] Operations were immediately suspended and hundreds of men were thrown out of work, with many even leaving the county in search of employment.

Although the injunction was destined to be short-lived, causing the *Transcript* to enthuse, "the people who a fort-night ago, were despondent, are now contented with the prospect of a prosperous winter."[152] Such rhetoric, however, proved to be mere "whistling in the dark." By September 1882, Edward Woodruff of New York, an absentee landlord with vast holdings of Yuba farmland, asked for a perpetual injunction against the North Bloomfield Gravel Mining Company and after a year and one-half of legal delays, Judge Lorenzo Sawyer of the United States Circuit Court, on January 23, 1884, issued his fateful determination. Within a matter of days the North Bloomfield Company ordered all men off their line of ditches and gave instructions to have the waste gates thrown wide open. This action included 157 miles of canals and flumes reckoned at a construction cost of $708,841. The Milton Company and the Eureka Lake Company soon followed suit—for all practical purposes large-scale hydraulic mining in the Sierra Nevada was over, forever.[153]

In August 1883, H. C. Perkins, for many years superintendent at

North Bloomfield, announced he was leaving for a similar position in South America. It was reported his salary would be $18,000 per year—a staggering amount in the 1880s. His good friend and associate, the distinguished engineer Hamilton Smith, Jr., also left for Venezuela the following month. Both of these men realized the futility of continuing the fight for hydraulic mining. Smith who was the founding president of the Miner's Association, and for years the most active leader in behalf of the miners, was also the preeminent hydraulic mining engineer of his era. During his tenure as engineer, general manager and later president of the North Bloomfield Gravel Mining Company, he obtained the background for his definitive treatise, *Hydraulics*. This work became recognized worldwide as the authority on hydraulic mining. His paper "Water Power With High Pressure and Wrought Iron Pipe," won him the coveted American Society of Civil Engineers prestigious Thomas Fitch Rowland Prize in 1884.[154]

Less than a month after the Sawyer Decision, the North Bloomfield Company published its financial statement for the period, January 1, 1883 to February 1, 1884.

Total receipts from bullion	$ 483,187
Total receipts from water	23,009
	$ 506,196
Paid out	
Redemption and interest on bonds	$ 223,778
Dividends	180,000
Operating expenses	170,106
Labor	87,309
Total bullion product until injunction	2,829,870
Total assets	$1,967,933 [155]

During the months and years following the Sawyer Decision, limited mining was conducted at the Malakoff and even drifting was attempted, but with minimal success. In 1888 it was reported that two monitors were in operation—one six-inch nozzle and one seven-inch. Also a Hendy elevator (Evans) was at the site with a vertical height of 90.6 feet on an incline of 60 degrees. The water was delivered with a volume of 1,400 miner's inches and a pressure of 530 feet. As late as 1895, author Kirk Monroe was so fascinated by the operation of the Evans elevator at the Malakoff mine, and so convinced that this device would be the salvation of the industry, he described it in glowing prose in an article for *Harpers Weekly*.

> They discharge such a mighty rush of water that from the top of the shute it makes a vertical leap of 200 feet, and falls to the ground 500 feet away. This is the hydraulic elevator, and by its irresistible stream all material brought down from the sluice, including paving stones as large as one's head, gravel, sand and clay, is lifted over the dam and flung to a distance beyond it. . . . So tremendous is the force of the hydraulic elevator that though the massive lumber shute is lined with plates of manganese steel two inches thick, these are worn to the thinness of paper in a few months by friction with the out-rushing gravel which is projected with a velocity of 176 feet per second and a roar that echoes for miles down the canyon.[156]

Following the passage of the Caminetti Act by Congress and the formation of the California Debris Commission in March 1893, small-scale hydraulic mining was resumed at North Bloomfield. However, with the increasing severity of restrictions passed by the commission, operations were finally brought to a halt in 1900.

Newspaper correspondent Samuel Butler visited the moribund town in 1893, recording his impressions of what once was the busiest mining camp on the San Juan Ridge.

> We proceeded to the town of North Bloomfield arriving there about 6:00 p.m. To say that we were gratified by the appearance of things would be a prevarication of the vilest kind. There was a dull monotony everywhere prevalent which was painful to behold. Bloomfield today is only a mere skeleton of what it was during the prosperous era of hydraulic mining. . . . The town is situated 14 miles northwest of Nevada City and is the largest town on the San Juan Ridge. The town is not very compactly built, the houses being scattered on account of the principal mines being some distance from the main portion of the town. The town is not devoid of attractions, for it contains a theater, Catholic and Methodist churches, and a well constructed schoolhouse.[157]

From 1866 to 1900, it was estimated that about 30,000,000 cubic yards had been worked in the vicinity of North Bloomfield, amount-

Hendy Giant on view at Malakoff Diggins State Park. Formerly used at the La Grange Mine, Trinity County.

ing to a total production of approximately $3,500,000. Certainly the ground at the Malakoff was far from the richest on the Ridge, but the scope of its gravel deposits were awesome. If allowed to continue in full production, it would in a few more years, possibly have become the greatest producer in the state. It was calculated by the State Mineralogist that 300,000,000 cubic yards of auriferous gravel still remained with an average value of ten cents per yard (1918 prices).[158]

All recent attempts to resume placer mining on the San Juan

Ridge, using the latest and most advanced techniques, have been vigorously and successfully opposed. Today, it appears, the treasure of an unaltered landscape is more highly esteemed than that which lies buried beneath the ground.

In 1966, exactly 100 years after the incorporation of the North Bloomfield Gravel Mining Company, the state of California created the Malakoff Diggins State Historic Park on 3,000 acres of land. It included the colorful Malakoff pit measuring 7,000 feet long and 3,000 feet wide with a depth of approximately 600 feet. The remaining buildings along the main street of old North Bloomfield have also been restored and are on exhibition. The collection of hydraulic mining equipment and other displays in Cummings' Hall, now the park's museum, together with several hydraulic monitors (including a Hendy Giant) lining the street outside, helps the visitor to better understand the extremely important, yet, turbulent and controversial story of hydraulic mining in California.

Chapter 6

The Ending

If the hydraulic method is to be indefinitely used, without restraint, upon all the surface that will yield a good return, the California of the future will be a waste, as though demons had inverted the Creators' intention in it—a waste more repulsive than any denounced in prophecy as the doom upon a guilty race.

Thomas Starr King, 1860

As early as 1856, the waters of the Bear and Yuba rivers, where they entered the Great Valley, were perceptively polluted from mining debris, causing concern for the future of the adjacent farmlands. But with the slowdown in mining in the late 1850s, it did not become a pressing issue. However, by 1860 a few objective observers predicted that, unless hydraulic mining was restrained, in years to come, the Sacramento River would not be navigable and the rich farmlands of its valley would become a vast wasteland. But even the farmers, who potentially would be most directly affected, showed little concern—that is until the great storm of the winter of 1861-62. Beginning December 8, six inches of rain fell at Nevada City in a period of 24 hours and by season's end the total rainfall was an alarming 115 inches. The streams and gulches of the Sierra foothills, which had been choked with years of accumulated mining debris, let go their oozy caches depositing tons of slickens on the farmlands adjacent to the Bear and Yuba rivers. While the anguished cries of the farmers were directed against the hydraulic miners, ironically, most of the tailings were the product of ten years of ground sluicing in the Sierra foothills.[1]

But the great mass of debris that would clog the rivers and streams flowing into the Sacramento and San Joaquin rivers in the following years was sired by hydraulic mining. In 1868 the beds of the Yuba and Feather rivers had actually become higher than the city streets. The vulnerability of the City of Marysville, even with a system of levees, was witnessed during the great flood of 1875 that washed over the artificial banks and flooded the town. The direct result was an act by the State Legislature creating a "Board of Levee Commissioners" who were empowered to purchase land, levy taxes and construct an elaborate system of levees, costing nearly $1,000,000 and virtually surrounding the City of Marysville so that even today it is known as "the walled city."[2]

For over 15 years the newspapers, whose constituency farmed the lands adjacent to the Yuba, Feather and Bear rivers, had alerted their readers to the increasing danger from the slickens pouring into the valley, but their protestations reached a very limited audience. The rest of the state was rather oblivious to their condition. It is true that a few articles antagonistic to the ravages of hydraulic mining found their way into national publications, but their concern focused mainly on the damage to the mountain environment rather than the farmlands below. Typical of this genre was a rather Victorian recital written by A. Judson Farley for the *Overland Monthly*. The poignant article began with an idyllic description of a site, later to be named Rose's Bar, located on the south bank of the Yuba River; but its beauty, the reader learned, was destined to be despoiled by years of hydraulic mining. The tale concluded with a description of the desolation that remained after the monitors had taken their toll.

> Before you are the hills, essentially the same, but marked with unmistakable signs of an energy which will not be satisfied. Unlike the hills, the Yuba, which used to sing and smile for them with a face pure and beautiful, forever ready to reflect their grace or strength, is now sadly changed and often furious. The white cataract has gone, and the banks which formerly

THE ENDING 229

This non-descript building houses the Marysville Levee Commission, established by an act of the California State Legislature in 1878.
Author's collection.

included it have almost wholly disappeared. Covering the hills is a patchwork of flumes and ditches, some of them deserted, but more of them noisy with the rush of the muddy and gold-laden waters. Here and there along the flume are small buildings, adapted for the purpose of hydraulic mining. The valley of Rose's Bar is no longer visible. In its stead is a long uneven bed of sand and cobble-stones interspersed with the cast-off clothing of the miners and the *debris* of his handiwork. Over this bed ran numberless small streams of muddy, yellow water, sent from the mouths of the wooden flumes above the yellower river below. Buried a hundred feet below is the course of that river, and buried with it too, is all that was of Rose's Bar. The flowers, fresh and sweet, of its earlier history; the footpaths of a later period, warn by the hardy men of the camp and leading down from it to their work in the stream close by.[3]

By the beginning of the decade of the 1870s, even the mining town newspapers began to take notice of the plight of their neighbors in the valley. The Nevada City *Daily National Gazette* acknowledged that in Nevada County alone, 100 hydraulic pipes were moving 100,000 tons of debris daily, totaling 30,000,000 tons annually. The combined hydraulic mines in the state were responsible for 90,000,000 tons of slickens each year, sufficient to cover over 987 square miles to a depth of one inch. While admitting that the quantity of debris moved by the hydraulic process in California was considered enormous, the newspaper consoled its readers by concluding, "but when we reduce it to a cube, it appears insignificant."[4]

As Nevada City was the center of the hydraulic mining interests in the Sierra foothills, so Marysville, located at the confluence of the Yuba and Feather rivers, was the focal point of anti-mining sentiment in the Sacramento Valley. Likewise, the Nevada City *Daily Transcript* and the Marysville *Appeal* became the leading organs espousing the two opposing viewpoints. The vitriolic rhetoric exchanged by the two disputing editors became almost a personal vendetta with comic overtones full of hyperbole and exaggeration.

Under the heading "Slickens For Potatoes" the *Transcript* facetiously responded to the town of Wheatland's continuing appeal for a cessation of hydraulic mining: "It is said that the soil deposited by the discharge of water from the mountain streams where mining has been done is what makes Wheatland so good a point for raising potatoes. This soil needs no irrigation and potatoes grow in it of the largest and best quality."[5]

The farmers near Wheatland, on the banks of the Bear River, were not amused. In the early spring of 1876, one of their number, James H. Keyes, entered an action in the Tenth District Court in Yuba City bearing the title, *Keyes vs. Little York Gold & Water Company et al*, which included 18 other mines that dumped their tailings into the Bear River. The *Transcript* it appears still could not take the farmers protest seriously. "The Wheatland farmers have concluded to sue the hydraulic miners whose tailings are dumped into the Bear River. Well let them sue and see if they have the right to stop the matter."[6] In late July the Sheriff of Sutter County appeared along the Bear River in Placer and Nevada counties serving summons on all hydraulic miners to appear in court to show by what right they could send slickens to the valley below. To cover the large legal expenses needed to carry on their fight, the farmers created an informal organization of sympathizers to pool their funds.[7]

With this turn of events the mine owners began to feel a real threat from the farmers and saw the need for immediate organization. Keyes had hired a prominent Sacramento attorney George Cadwalader who would prove a thorn in their side for years to come. As early as the following July, a notice was mailed to all hydraulic gravel mine owners and canal company owners advising them of the need for the formation of the Hydraulic Miner's Association "for mutual benefit and protection," signed by Hamilton Smith, Jr. "Chairman, Temporary Organization."[8]

By then the *Transcript* had switched to a confrontational take-it-or-leave-it hard line.

> The channels of Bear River and the canyons leading into it have been filled in places to a height of 70 or 75 feet by the light topsoil which has been washed off from hundreds of acres of land. . . Bear River, from the edge of the valley to Dutch Flat, stands full of it. Great trees which once stood on the banks of the clear and swift mountain stream now are buried to half their height, and a mass of bushy limbs, always dead or dying projects from the beds of white sand and rocks. . . . So far as the deposit of gold-bearing dirt is concerned it will last longer than the mines can be operated. There are miles and miles of pay still unopened. The end will come when the beds and canyons fill up to the mouths of the different tunnels, which will be several generations for the best mines unless the law interferes.[9]

In the meantime, the farmers had succeeded in getting a change of venue to the Federal Courts in the Keyes matter, but a period of two years would pass before a decision would be announced. In the interim, the State Legislature was not idle. Yielding to the pleas of the farmers, the so-called Haymond Bill was passed creating a state Engineering Office to examine the debris question and to report back to the 1880 session of the legislature with a plan that could be pursued for a solution to the problems. For months prior to this action, the Grass Valley *Union* had been admonishing the hydraulic miners that the residents of the upper Sacramento Valley had active agents hard at work in the State Legislature warning that "the miners pour water against the gravel and do not seem to care a cent what sort of a stream of legislation may soon be turned against the great interest of hydraulic mining."[10] As early as September 1879, the new State Engineer, William Hammond Hall, had begun his survey. In this endeavor he was accompanied by G. F. Allardt, representing the farmers and Colonel A. W. Von Schmidt, a consulting engineer, representing the Hydraulic Miner's Association. Although he did not accompany the survey party, the Association had also appointed Walter A. Skidmore to act as its special agent and representative in the controversy with the farmers.

The following month, in an effort to attract a broader base into the association and include the entire mining industry, the name of the organization was changed from Hydraulic Miner's Association to simply, Miner's Association. The new officers were listed as Hamilton Smith, president; L. L. Robinson, vice president; and Lazard Freres, treasurer. Their offices were located in San Francisco at 320 Sansome Street, Room 23 (the same as the North Bloomfield Co.).[11]

In January 1878, the United States Supreme Court decided the Keyes case was not a matter for the Federal Courts and returned it to the District Court in Yuba City. A little over a year later, March 1879, the long-awaited decision was announced, with the court handing down a permanent injunction preventing miners from discharging their tailings into the Bear River. The miners immediately filed an appeal and the injunction was temporarily lifted, allowing them to continue to operate. Judge Keyser's court order went into effect on June 17, 1879. Sargent & Jacob's claim at Quaker Hill was the only mine to close down. Commenting on the matter, the *Transcript* complained:

> The great slickens case has at last been decided so far as Judge Keyser's authority in the premises is concerned. After several months testimony and argument introduced during the trial of the matter of Keyes vs. Little York Mining and Water Company et al, last summer, in the Tenth District Court at Yuba City, he rendered a judgment Monday that was in strict accordance with what might have been expected to emanate from a court elected on the issue as to sending down of debris should be stopped or not. He has granted a perpetual injunction in favor of the plaintiff . . . the plaintiff, a farmer residing on the Bear River in Sutter County. . . . We cannot believe that the Supreme Court will sustain Keyes' decision.[12]

The example of the organization of the hydraulic mine owners, together with the expenses of the long drawn-out Keyes case, convinced the farmers of the pressing need for a more formal association. On August 24, 1878, a meeting was called at the Sutter County courthouse in Yuba City. On that day "The Anti-Debris Association of Sacramento Valley" was formed for the expressed purpose "to prosecute to a final decision any pending or future case which tested the right of the miners to use the rivers for deposit."[13] In less than a month's time the city of Marysville filed an action against the North Bloomfield Gravel Mining Company in the Yuba District Court and the battle lines seemed clearly drawn.

The farmer's position, with respect to the legal issues involved, was clearly expressed by the editor of the *Appeal*: " Every lawyer and every judge knows that as between the rights of citizens the miner commits a trespass and wrong when he sluices sand and rocks upon the agriculturists and converts his fertile lands into desert waste. The act is so preposterously wrong that we at once say there can be no relief for the hydraulic miner in the Supreme Court of this state."[14] But the next round was won by the miners when in November 1879, the State Supreme Court invalidated the injunction issued in the Keyes case. This important decision was based on the principle that one can not be held accountable for the wrong which he is not a part of merely by joining him with another defendant who may have independently threatened a similar wrong. In legal terms this canon was known as a misjoinder. The reader may recall this was a complete reversal of the principle used in settling the suit against the Spring Valley Company by the farmer named Crum.

On November 18, the *Transcript*, ran the following headline:

VICTORY
THE STATE SUPREME COURT GETS DOWN TO ITS KNITTING
AND FIRES THE SLICKENS CASE OUT THE WINDOW

The column then explained to its readers that if the farmers intend to prosecute the miners they must take one company or individual at a time instead of bunching them all together. The news of the lifting of the injunction was joyously celebrated in the mountain towns with "flying flags, booming bells, crashing cannons, shouts and the blowing of steam whistles."[15]

Early the following year, the long-anticipated report on irrigation and mining debris, prepared by State Engineer, William H. Hall, was presented to the State Legislature. The grave situation and the potential disaster that would assuredly follow was outlined in solemn detail to the state's lawmakers. As a result of this alarming report and after a clamorous uphill fight, mainly sectional in nature, the so-called Drainage Act was finally adopted, April 9, 1880. The new measure would permit hydraulic mining to continue, but only after adequate precautions were provided to protect the farmers. Three commissioners were to be appointed by the governor to administer the program. The state was to be divided into drainage districts and the State Engineer would oversee the construction of impounding dams and a series of levees.

By early July, Engineer Hall submitted plans to the Drainage Commissioners for two debris dams to be constructed on the Yuba River and another on the Bear. One of the Yuba dams was to be located at a site about ten miles above Marysville known as Daguerer Point, but was never constructed. The other, after an initial change, would be constructed at an extremely wide location on the river about two miles below Daguerer Point. The following month more details about the actual fabrication were available. The upper dam on the Yuba would stretch across the riverbed for a distance of 4,800 feet and rise to a height of eight feet, composed entirely of horizontal layers of trees from four to eight inches in diameter and from 25 to 30 feet long. The trees, with the brush still on their trunks, were to be placed with their butts downstream. In this manner, in a short time, they would become filled with gravel and silt, making a satisfactory impounding dam. The Bear River Dam was engineered and constructed in exactly the same fashion. The lower Yuba dam was of similar construction, but because of the exceptional width of the river at that point, although never fully completed, it was designed to be over 9,000 feet in length.[16]

For months the hydraulic miners had argued that legislation,

THE ENDING

rather than litigation, was the proper way to settle the debris controversy, and with the enactment of the Drainage Act they presumed the farmers had also, albeit reluctantly, arrived at somewhat the same position. From subsequent events, the miners would become painfully aware their assumption had little basis in fact.

Within a matter of weeks after the enactment of the Drainage Act, the United States Congress, out of concern for the navigable rivers and bays in California, asked the Secretary of War to explore these vital waterways to determine if they were, indeed, being affected by mining debris. As a consequence, Lieutenant Colonel G. H. Mendell, of the United States Corps of Engineers, was ordered to make an investigation and report back with appropriate recommendations.

In May of the following year the city of Marysville, without warning if the anguished protestations of the *Transcript* were believed, revived its suit against the North Bloomfield Gravel Mining Company. Two months later an injunction was served on them and their affiliate, the Milton Mining & Water Company and all other hydraulic mines that emptied their tailings into the Yuba River. This action by Marysville, purportedly, came as a complete surprise to the miners who still claimed they had arrived at a compromise with the farmers. The *Transcript* noted, ". . . on account of litigation which has been started against hydraulic mining, most of them [hydraulic mines] have suspended operations." In retaliation against the city of Marysville, for what was proclaimed a wholly unwarranted action, the miners and merchants in the mountains, and particularly in Nevada City, began a campaign to boycott merchants and other businesses in Marysville hoping, out of economic necessity, it would influence them to side with the miners. This action prompted the San Francisco *Daily Exchange* to write:

> There is one thing that the gravel miners seem determined upon and that is to hold the city of Marysville directly responsible for the stoppage of their operations. That place by unanimous consent is being rigidly "boycotted." Her merchants cannot sell a dollar's worth. Her farmers seek in vain a market for their produce. Debts are uncollectable and a deadly hostil-

ity exists between the two sections.... The *Exchange* has no hesitation in declaring that its sympathies lie wholly with the miners. We believe that the crusade against the miners began without warning by the people of the valley, and carried on without regard to justice or equity or the rights of others and with a vindictiveness wholly uncalled for, is an outrage upon the victims, which justifies "boycotting" or any other manner of redress.[17]

Although the Marysville *Appeal* repeatedly insisted that the boycott was having little if any effect on the local economy, other sources indicate this was not exactly the case. Making conditions even worse in town, the city fathers passed an ordinance closing all businesses on Sunday. As the Marysville *Express* put it: "What is to become of Marysville? From the agitation of two leading questions we can see the business of our merchants and public houses steadily declining day by day.... The instituting a suit in the name of the city had the opposite effect of increasing trade...."[18]

In a series of letters written from Smartsville, by a party signing himself "Comet," and published in the *Appeal*, the hydraulic miners were assailed in a "dastardly and lying way," according to the editor of the Nevada City *Daily Transcript*. "They urge the grangers on to renewed activity in the matter of trying to close down the mines. The author is no other than the universally despised store keeper at Smartsville, whom the miners there, for obvious reasons, will not patronize.... We are surprised that the *Appeal* should permit him to use it as a means of vengeance."[19]

One interesting ramification to the boycott of Marysville was the situation with respect to the Empire Foundry. The reader will recall that it was owned by Richard Hoskins who manufactured the Little Giant monitor and enjoyed a monopoly in that area. The fact that these "engines of destruction" were manufactured in Marysville and sold to the miners could be compared to a munitions maker selling arms to the enemy, but the people in that town didn't look at it that way. They reasoned that someone would be making them elsewhere so it might as well be in Marysville—at least it provided a living for a number of families. It was also rationalized that most of the monitors were being manufactured for customers in other states or overseas. But to the editor of the *Transcript*, the manufacture of hydraulic monitors by the Marysville foundry, a product born and bred in Nevada City, was a source of extreme irritation. On a number of occasions he indicated that the business should be returned to Nevada County—even suggesting that since the patent rights had expired on the Little Giant (they really had not), the Nevada County foundries should start manufacturing monitors on their own account, adding ". . . if Marysville wants to try and ruin us we must by all means retaliate."[20]

Much of the agitation in Marysville stemmed from the fact that the impounding dams had not held the debris during the storms of the previous winter—even though the construction of the upper dam on the Yuba was never begun and the lower dam was never fully completed. Here again, the Marysville *Express* took a moderate tone. "It was not the miner's fault if the dams did not, during last winter, fully answer the purpose for which they were designated. They were incomplete. . . . Yet crippled as they were, no one can deny that they have to a very great extent answered their purpose. . . . the Yuba dam is pretty well filled and the millions of yards of debris confined within it would certainly . . . have come down into the lower portion of the river."[21]

During the period of mining closures, the leading hydraulic mine owners in Nevada County were asking the County Tax Assessor to reduce the assessed valuation on their property by 66-2/3 percent. Daniel Collins, the County Assessor, stated that in his opinion, if hydraulic mining was stopped, there would be a reduction of half the assessment rolls from nine million dollars to four and one-half million dollars and would wipe out the whole of the property of Bridgeport, Bloomfield and Eureka townships.[22]

On June 12, 1881, an important notice appeared in all of the newspapers of Nevada County and of those in most of the other hydraulic mining counties as well. A meeting of hydraulic mining interests was to be held at the Nevada City Theater the following Saturday at 7:30, June 18. Included in the notice was a long list of important men in the mining industry who would be in attendance.

It was known that the mine owners were in desperate need of financial help and that a strong appeal would be made for general support. Subsequent events would prove that the rank and file miners also had an agenda of their own.

A huge crowd filed into the building from all of the hydraulic mining counties and the meeting was called to order by the temporary chairman, the Honorable W. D. Long, who announced that the purpose of the meeting was to "prevent, if possible, a dire calamity." After the election of officers, a resolution was approved to support the Drainage Act passed by the State Legislature. This was followed by a number of other resolutions, and concluded with money-raising pep talks by several of the leading hydraulic mining men in attendance. But as far as the miners were concerned, the most important bit of business accomplished at the meeting was a resolution they initiated, calling for the end of the practice of hiring Chinese labor. Included in the argument was the statement—if hydraulic mines could not be operated profitably without Chinese, then shut them down. With the mine owners pleading for financial support from their own labor force, they had no option but to readily accede.[23]

Knowing the *Transcript's* rabid position on the Chinese question and the apparent failure of the boycott, the *Appeal* never missed an opportunity to probe a wound.

> Marysville will not be materially injured by a withdrawal of the trade from the mining camps. The dealers there do not get more than ten or fifteen percent of their patronage from the hydraulic mines. . . . Since the large hydraulic mines at Smartsville shut down the Chinamen in that neighborhood have taken advantage of the opportunity to buy water in unlimited quantity and are mining day and night. They formerly only worked in the daytime. Their operations are conducted on a small scale, but they are making the best use of their time. Real estate has taken a big tumble since the mines stopped working. Most of the unmarried miners have gone away to look for work.[24]

In the latter part of April 1881, a suit was filed in the Superior Court in Sacramento to test the constitutionality of the Drainage

Act and, in the following September, the State Supreme Court declared it unconstitutional, because it concerned itself with more than one subject and those subjects were not contained in the title as required by the constitution.

Although W. F. Knox, one of the Drainage District directors, had pronounced the Drainage Act a failure, he did admit that it had accomplished one good thing. It had, at last, brought to the attention of the entire state, the magnitude of the destruction from the mining debris and the urgent necessity of doing something about it. In his opinion, if these facts had been known ten years earlier, millions of dollars might have been saved. He also noted that those outside the affected area paid little attention until their pockets were touched.[25]

Remarking on the demise of the Drainage Act, the San Francisco *Alta* took the position that it was not the business of the state at large to pay for the proposed dams and reservoirs, because they appeared to be mainly in the interest of the "swamp-land speculators," and because the money raised by taxation seemed to have been wasted on a couple of worthless brush dams.

The following month the same court ruled that Judge Phil W. Keyser was not competent to issue injunctions in the case of *Marysville vs. The North Bloomfield Gravel Mining Company et al.* This decision was based on the simple fact that the judge owned land in Marysville and, therefore, there was a conflict of interest. When the news hit the streets of Nevada City, there was wild rejoicing. On June 18, The *Transcript* displayed a large headline.

CARRY THE NEWS!
A KNOCKDOWN FOR THE MINERS
THE SUPREME COURT SITS DOWN ON KEYSER
THE INJUNCTION ON THE
HYDRAULIC MINERS RAISED

But in the same edition, editor L. S. Calkins had some rather scathing words for unspecified citizens in town who, he felt, were not supporting the Marysville boycott as effectively as they might. "Judas Iscariot sold out Jesus Christ for thirty pieces of silver. It is an open question if there is a man in Nevada City who would dispose of himself at such an insignificant price. It hardly seems possible there can be one, but a short time will determine the fact one way or the other. During the struggle now going on between Marysville and the miners, it is the bounden duty of everyman . . . to take a personal stand upon the all-important debris question."

As a result of the rather complete breakdown in any form of negotiations between the two opposing sides, the San Francisco Board of Trade formed a Debris Committee to launch an investigation and make recommendations to resolve the impasse.

For some time past the farmers of the rural counties of the Sacramento Valley had been exerting pressure on State Attorney General A. L. Hart to bring suit against the hydraulic mines that dumped their tailings into the Feather River. As a result the Miocene Mining Company of Butte County was targeted for the action. However, when the New York-based owners insisted that the case be moved to the federal courts, Hart backed off. Seeing a window of opportunity, the Sacramento Board of Supervisors petitioned Hart to transfer his action to one of the mines depositing debris into the American River—which he agreed to do. As a result, in July 1881, an injunction was requested against The Gold Run Ditch & Mining Company, which appeared to be the mine of choice.[26]

In the meantime, Judge H. A. Mayhew of the Tehama County Superior Court reissued the injunction against the North Bloomfield and Milton companies in the matter of the *City of Marysville vs. the North Bloomfield Gravel Mining Company*. As a consequence Sheriff Tomkins was ordered to "Shut the water off." This precipitous action caused the North Bloomfield Company to immediately obtain an injunction issued by the Nevada County Superior Court, restraining the sheriff from trespassing on the mining lands for the purpose of shutting off the water.

Judge Mayhew's injunction did, however, close a great number of the Yuba mines, many in Yuba County near Smartsville—not too distant from Marysville. For this reason the *Transcript* enjoyed featuring the pain and suffering Marysville was causing its own people. The employees of the Excelsior Water & Mining Company had been notified early in June that the mine would be shut down the next day because of the injunction. "The foreman and all the workers were all discharged," the paper explained, "several hundred men, a majority of whom have large families to support, will be obliged to seek other homes unless mining is resumed. The loss to the laboring class will be $25,000 per month."[27]

Meanwhile the boycott continued and the bitterness toward the City of Marysville intensified. In early June, two wagons from Marysville, one loaded with potatoes and the other with hay, arrived in North Bloomfield to sell their wares. Needless to say they were met with hostility and ordered to leave town. Two days later, the North Bloomfield mine suspended operations because of the restraining order. The following week the *Transcript* noted that the mine had invested about $2,000,000 in Nevada County, but now they employed only about ten men to take care of the property and soon expect to reduce that number.[28]

In still another answer to the *Transcript*'s continuing claims of the effectiveness of the Marysville embargo, the *Appeal* filed the following story:

> If there has been any interruption of traffic between the mountain towns and this city, that traffic has been fully resumed. Yesterday there seemed to be more large freight teams moving about the city than has been seen any day this season. In answer to an inquiry Mr. Stratton said that he had yesterday shipped goods to Camptonville, Dry Creek, French Corral and Strawberry Valley. He added that the cessation of hydraulic mining in some camps has lessened the demand for some kinds of merchandise but he did not know that any boycotting was going on. Mr. White of the firm of White Cooley & Cutts, remarked that a mountain trader, who ten days ago gave notice that he would buy no more goods in Marysville, came in yesterday and bought a supply of goods. "You see," said Mr. White, with a good natured smile, "a man is not going elsewhere to pay a dollar and a half for a thing that will cost him only a dollar here."[29]

As the debris controversy heated up, the Marysville *Appeal* was joined by a strong ally in the form of the powerful Sacramento *Bee*. In a July issue in 1881, the newspaper ran a lithographic cartoon produced, in pencil, by the well-known cartoonist G. F. Keller of the San Francisco *Wasp*. The viewer beheld a large hydraulic mine with powerful streams of water being forced against a bank in full blast. A few Chinese were operating the monitors (the anti-debris people insisted that large forces of Chinese worked in the hydraulic mines). The water was tearing away the earth, rocks and trees from the face of the hill, sending it all down in an avalanche of mud. At the foot of the mountain farm houses were seen half covered in sand and fruit orchards were nearly submerged, with threshing machines and other farm implements buried and abandoned. Live stock were also shown drowning in the floods that stretched for miles down the Sacramento Valley. In the foreground was a farmer with his family mounted in their wagon, packed with all their worldly possessions having to abandon their home. Standing on the side of the mountain were a couple of hydraulic mine owners, fat, sleek and well dressed, with fat cigars, complacently viewing the desolation.[30]

The subject of the cartoon was prompted by a story in the *Appeal*, describing the heartache of 12 families near Wheatland, who actually abandoned their farms, ruined by slickens, to move to Texas and start a new life.

The farmers in the valley often expressed their impatience with the hydraulic miners for their unwillingness to end the debris problem by simply shifting to drift mining. Everyone knew that both methods could mine Tertiary gravel deposits—so why not? In reply to these demands the *Transcript* explained that the cost for drifting ran, on average, from $1.00 to $2.00 per cubic yard. The column then listed the yield from a number of hydraulic mines to prove how impossible such a switch would be: The American Mine, 24 cents per cubic yard; Blue Tent Mine, 15 cents; North Bloomfield Mine, less than 10 cents; Milton Mine, 33 to 47 cents; Manzanita Mine at Sweetland, 27 to 33 cents. Supporting this view, Lindgren would later write: ". . . The gravel outside the pay streak [in a drift mine] would ordinarily be regarded as extremely rich by the hydraulic miner who would be content with a yield of 10 cents a cubic yard; but the drift miner is obliged to leave as unpayable gravel containing from 75 cents to $2.00 per cubic yard." Taliesin Evans reported that six prominent hydraulic mines, over a period of two seasons, showed an average yield of just seven and one-quarter cents per cubic yard.[31]

In early November 1881, the Debris Committee of the San Francisco Board of Trade, acting as arbiters, submitted their recommendations to resolve the impasse and effect a compromise. They proposed that a stone dam should be immediately constructed on the Yuba River below the junction with Deer Creek, possibly at Landers Bar, and at other locations. Also, they recommended that the brush dams on the Yuba should be repaired. The entire cost would be borne by the mining interests and the Federal Government was urged to appropriate enough money to dredge the Sacramento and Feather rivers in the interests of commerce. The committee also recommended that all litigation between the parties be suspended. Finally, it was proposed that a convention be called in San Francisco to convene on Thursday the 17th day of November. It was to be composed of 15 delegates from both the Anti-Debris Association and the Miner's Association and, if advisable, ten disinterested citizens from the Bay Area should be selected to also serve as members.[32] Not surprisingly, the Miner's Association informally agreed to the board's proposal, but the farmers steadfastly declared that the only way they would end litigation was when hydraulic mining was forever prohibited.

The following year, in early February, Colonel Mendell's report and proposals were made known to the anxious miners and expectant farmers of Northern California. His recommendations, which called for an appropriation of a little over $500,000 to be used for the construction of dams on the Yuba, Bear and the American rivers, came as a relief to the miners, but consternation and alarm to the farmers—they had hoped Mendell's report would spell the death of hydraulic mining. His findings, however, were quite alarming. On the Yuba and Bear, the riverbeds were now higher than the adjoining countryside (no surprise) and from 1855 to 1873 the San Francisco Bar had suffered no injury. But from 1855 to 1878, the ship channel in San Pablo Bay had narrowed about 20 percent. During the period, 1867 to 1878 a deposit of 2,000,000 cubic yards had been found in the lower three and one-half miles of the Sacramento River and 500,000 cubic yards in the San Joaquin. Since 1867 shoals had increased in Suisun Bay.

Also, in recent years, deposits had formed in the Carquinez Strait. Near the mines, the Bear River had been filled some 250 feet, and the Greenhorn Creek, 200 feet. In 1878 it was estimated that there were deposited in the Bear and its tributaries 122,000,000 cubic yards of debris. In 1879 it was estimated that 72,000,000 cubic yards had been deposited in the Yuba River, not counting its tributaries. There had been found 20,000,000 to 25,000,000 cubic yards in the North Fork of the American above the junction with the Middle Fork. From the canyon of the Sacramento River, 6,000 acres of land were covered and the riverbed had been raised anywhere from five to 30 feet. In 1880 the State Engineer had estimated that 15,220 acres of land on the Yuba were seriously injured by deposits. Finally, the bed of the Feather River had been raised 15 feet at Marysville and five feet at its mouth.[33]

The rejection of the Mendell Report by the farmers was considered a crucial turn of events by the Miner's Association. Writing years later, its president, J. H. Neff, recalled their position: "It may be mentioned, as illustrating the importance of the issue, and of the first cause of this association, that the federal engineers estimated that, by spending $500,000, there could be $335,000,000 extracted by hydraulic mining in the region considered, without damage to the valley interests."[34]

"There will be trouble if the Blue Tent mining company does not cease giving employment to the small army of Chinamen working in its gold mines," warned the editor of the *Transcript*.

> We ask that company, how can it expect sympathy and financial aid from our people when it daily insults them by employing rice eaters in the place of white men? It was understood fairly and squarely, that no Chinamen were to be employed in that class of mining, and for that reason our

The Greenhorn Creek still shows the ravages of the hydraulic mining era. Photo taken in You Bet area. Author's collection.

people responded nobly to their call. While all the other companies have acted honorably, this company refuses to discharge its heathens. It is an outrage upon this community, and is condemned by all men of honor. The enemies of hydraulic mining have continually thrown in our teeth the assertion "that our hydraulic mines were owned by foreigners and therefore should not receive the support of Americans." This is untrue except for the Blue Tent Company. All other mines owned and controlled by Americans have discharged forever the Chinamen.[35]

The superintendent of the Blue Tent responded immediately, stating that the Chinese were only working during a time of "dead work," waiting for the completion of a bedrock tunnel and they would be dismissed as soon as it was completed. The superintendent explained that the Chinese were the only persons who would work week after week, getting their board only and waiting till the new bedrock tunnel was completed and other dead work done before receiving the remainder of the wages due them.[36]

On a happier note the *Transcript* was able to report the following month that ". . . the North Bloomfield hydraulic company are working about 150 men, all strapping, intelligent Caucasians, and most of them have families. About 25 of them are at work on the debris dam in Humbug canyon, and they will not get the structure entirely completed for several weeks yet."

In June 1882, Judge Jackson Temple, who had been deliberating the Gold Run Case, rendered his decision. A perpetual injunction was issued prohibiting that mine from discharging its "course" debris into the North Fork of the American River. While a heavy gloom spread over the hydraulic mining regions, the citizens of the valley were overjoyed. However, the *Mining and Scientific Press* expressed the opinion that perhaps their celebrations were premature.

> On the announcement of the debris decision by Judge Temple, those who only read the "head-lines" in the daily papers imagined that the whole question had finally been settled in favor of the farmers, and that hydraulic mining in California would be likely to come to a stop. THE MINING AND SCIENTIFIC PRESS at the time expressed the opinion that the decision was by no means as sweeping as was generally supposed, and that so far from stopping hydraulic mining, it gave the industry the opportunity, under certain restrictions, to continue under cover of the law. The real thing the decision did was to assert the principle that the miners had no right, as they supposed they had, to deposit their tailing in the bed of the stream. Now that time has elapsed. . . . The farmers do not take it as well as at first, and the miners take it the better, though they will be put to great expense in carrying out the provisions. . . . According to the decision rendered by Judge Temple, hydraulic miners will be allowed to run provided some means are taken by which the heavier and courser particles of the tailings can be impounded and kept out of the rivers.[37]

As a consequence of Judge Temple's decision, together with questions arising as a result of the Mendell report and also to settle some confusion among various mine owners, who appeared to be working at cross purposes, a convention of miners "on the Yuba watershed" was held at the Nevada City Theater on July 22, 1882. The meeting was well attended with over 200 men filing into the auditorium. The main purpose of the gathering was to get all mine owners

THE ENDING

to abide by the ruling of Judge Temple and, in association with other miners, build appropriate debris dams to prevent their tailings from flowing into the Yuba River. In a resolution, the assembly pledged that, with the aid proposed by the General Government, it was within the power of the miners, on the Yuba watershed, to prevent further damage by constructing one or more dams across the Yuba River for the purpose of impounding debris. But the mine owners disagreed with the Mendell report with respect to the filling of the San Pablo and San Francisco bays. It was the opinion of the assembly that this was largely from agricultural and natural causes and that it was unjust to place any such damage solely upon mining.[38]

Almost two months to the day, after the meeting of the hydraulic mining leaders at Nevada City, the *Transcript*, under the heading "More Slickens Suits," ran a brief news account: "A suit has been brought in the United States Circuit Court by one Woodruff, against the Excelsior, Milton, Bloomfield, Sailor Flat, Omega and Manzanita mining companies, and the United States Marshal or his deputies may be expected to serve injunctions at any moment." From Smartsville to the San Juan Ridge, the same questions were asked—who was this Woodruff anyhow? How come he didn't include in his suit the Eureka Lake, Blue Tent and 20 or 30 other companies? Why had Woodruff only singled out five of the companies? Did he think that five at one pop would do to start with? Would he or the U. S. Marshal succeed in injuncting of these five companies?

Two days later an article in the *Transcript* shed a little more light. The United States Circuit Court Judge who was responsible for the papers to be issued in the case of *Woodruff vs. North Bloomfield Co. et al*, was no one else but the Honorable Lorenzo Sawyer and old resident of Nevada City (the man who described the coyote diggings in the first chapter). The story then revealed: "He is one of the most distinguished jurists in the state and the miners are well satisfied to rest their case in his hands, but we presume the anti-miners will commence kicking when they find out that he is not a thoroughbred granger. Whatever Judge Sawyer's decision may be, the anti-miners will not be satisfied, as they never are unless a judge says point blank all hydraulic mining must be stopped." If ever an editor wrote an article he later wished he could erase from the record—this was it!

Before the week was out, the name Edward Woodruff was on everybody's lips. It was learned he was a citizen of New York, who not only owned a block of stores in Marysville, but also an undivided part of the Hock farm tract (Captain John Sutter's farm) on one side of the Feather River and of the Elizabeth tract on the other side, both well-known, productive parts of the old Sutter grant. The complaint he entered stated that the mines named in the suit dumped 60,000 cubic yards daily, using 13,000 miner's inches of water. It was also learned that at least 30,000 tons of grain were shipped annually down the Feather River to market, but if the dumping of debris was not stopped immediately, navigation on the Feather and Sacramento rivers would be completely stopped. It was also stated in the argument that grain barges filled only to one-half of their capacity could barely get from the Feather to the Sacramento River by means of the Government snag boat *Seizer*. The Woodruff property in Marysville was only made habitable by a system of levees, which required a levee tax of from three to four percent per year, to save his land from inundation. It was becoming painfully obvious to the miners that Edward Woodruff had not just ridden into Marysville on a pumpkin wagon—they were in for a long hard battle.[39]

After the first phase of the hearings were under way, the *Mining and Scientific Press* added a somber note.

> The manner in which the case is brought, and the court in which it is to be decided, make this, in the estimate of the anti-debris men, the most important phase of the noted question ever presented in court.
>
> The importance of the case is suggested by the legal and business representatives of the two sides in attendance. The plaintiff is represented, legally by Judge Cadwalader, of Sacramento, Judge Belcher of Marysville, and Professor Pomeroy of San Francisco. The farmers are also represented by George Ohleyer, President of Sacramento Valley Anti-Debris Association, by ex-Mayor Bingham of Marysville, and a number of other promi-

nent Marysville and valley men. The defendants are represented by S. M. Wilson, Judge Wallace, I. S. Belcher and J. Byrne, counsel, and J. Hamilton Smith, President of the Gravel Miner's Association, Mr. W. A. Skidmore, Secretary of the Association, Superintendent J. O'Brien, of Smartsville, and many other mining men have been present in court. A large number of affidavits have been filed. A large portion of the time so far has been taken up in describing the customs of the miners.[40]

As part of his investigation, Judge Sawyer, with representatives from both the Miner's Association and the Anti-Debris Association, took a trip into the foothills to actually view the mines and, particularly, their restraining dams. Included in the party were L. L. Robinson, president of the Miner's Association (Hamilton Smith had earlier resigned this position); George Cadwalader, attorney for the Anti-Debris Association; James McClatchy, editor of the Sacramento *Bee*; Judge Niles Searls, representing the miners; and L. S. Calkins, editor of the *Transcript*. The junket, after leaving Marysville, traveled to North Bloomfield (via Nevada City), Columbia Hill, North San Juan, French Corral, then returned to their start via Grass Valley.

When the party arrived at the Malakoff near evening, they were met by Superintendent Perkins and were invited to dinner at his home. Upon visiting the mine that evening they found three monitors in operation with six-inch nozzles. The mine was well illuminated with electric lights and 125 employees were at work. The next morning they went to Humbug Canyon to view the impounding dam. It was found to be 52 feet high, backing up tailings for one-half a mile. However, it must not have been a very impressive sight for it had not been raised for several months, and debris was seen pouring over the structure. Leaving the Malakoff, the party went to North Bloomfield and the Derbec Drift Mine. From there they visited Moores Flat where they found the Snow Point and the Boston claims closed awaiting the Sawyer Decision. The following day they arrived at Columbia Hill then traveled on to North San Juan, and French Corral. The American Mine was not in operation, but both the Manzanita and the French Corral were running. The last night of the trip was spent in Grass Valley via Bridgeport, then back to Marysville the following day.[41]

James McClatchy, upon returning to Sacramento wrote an editorial for the *Bee* summarizing, in a list, his impressions gained as a member of Judge Sawyer's party, stating:

> 1. That hydraulic mining is but in its infancy.
>
> 2. That if long conducted as now the Sacramento Valley will be rendered uninhabitable.
>
> 3. That all the dams that have been put in or are likely to be constructed, to check the flow of debris into the valley, cannot hold back but a spoonful out of a bucketful.
>
> 4. That one or two things must soon come to pass — either hydraulic mining must be stopped or this valley must be deserted by men.[42]

At 5:00 A.M. on a Wednesday, June 18, 1883, a tragic event occurred that succeeded in changing the news focus of the contending editors, at least for a time. On that memorable morning the Rudyard Reservoir, commonly called the English Dam, gave way, causing a wall of water, 100 feet above the low water mark, to pour down the channel of the Middle Yuba River, sweeping everything before it, uprooting giant trees and removing ponderous boulders. N. C. Miller, the Milton ditch superintendent, shortly after the break, using the Ridge Telephone system, alerted all 22 stations. "The English dam broke this morning at 5:00 o'clock. Warn everyone along the middle Yuba River!" George Davis, the only eyewitness, who arrived at the site shortly after the break, reported: "It started by carrying off the wooden upper portion and then gradually crumbing the rest, stones and all until nothing was left but the site. The water was an hour and a half running out."

A house and barn at Jackson Ranch in the valley, three miles below the dam, were taken by the huge wave. The two Black brothers, who operated the ranch, crawled out of bed a little early that morning in time to see the flood approaching and managed to scramble to higher ground above its reach. A number of sheep that were grazing nearby were not as fortunate and were lost in the flood. At a point about one mile above the town of Eureka South (Graniteville), one of the two Reece brothers, "a quite decrepit" old man was swallowed up in the flood. First reports indicated that four Chinese and an Italian miner were also drowned at this location, but the story was never confirmed. At the McKillicum's mine (some reports call it Killican's) near Snow Point, the miners working in the tunnel would have been drowned by the flooding waters, but managed to escape by an air vent that had just been recently installed

The Eureka Lake flume, 12 miles above North San Juan, was carried away and at Emory's Crossing, a mile north of Freeman's, ten or twelve Chinese cabins were also lost. Four of the Chinese living there were drowned in the raging waters.

Freeman's Toll Bridge, 40 miles below the dam, located between Camptonville and North San Juan, was swept away in just minutes after the stage had crossed over it. The bridge, actually the fifth to cross the river at that site, was a covered structure 240 feet long, constructed in four spans, resting on piers rising 15 feet above the water. The flood arrived shortly after 9:00 a.m. and although the hotel and the Freeman house were saved by the water rushing around and through them, three or four small cabins, the blacksmith shop, with a 200 pound anvil, plus three outbuildings, one described as a broom factory containing a large quantity of brooms awaiting shipment, disappeared. Just before one of the cabins at the bridge was swept away, one Chinese man, despite warnings, ran into the cabin to rescue some of his personal property and lost his life. Freeman, himself, received word by telephone about 8:00 a.m. of the coming flood and watched the devastation from the safety of an elevated position. On the road from Freeman's, another covered bridge. located at Oregon Creek, was lifted off its foundation, swung eight or ten feet out of place, finally coming to rest on an elevated piece of land.

One of the passengers on the stage that crossed Freeman's bridge, minutes before the flood arrived, was Judge O. P. Stidger, editor of the *San Juan Times,* who penned a gripping account:

> Before the stage coach reached a point a quarter of a mile above Freeman's house, the roaring of the mighty waters gave notice that the flood

was close at hand. The stage had then reached a point where the passangers had a full view of the river, the bridge and Mr. Feeeman's houses. In a few minutes the flood came in all its fury, pushing, tossing and whirling in advance a great black mass composed of trees, logs, and driftwood of every kind and description with a perpendicular height of at least 30 or 40 feet. Viewed from our standpoint, it was the grandest as well as the most terrific sight the mind of man can concieve of. The water roared like thunder and moved at the rate of at least ten miles per hour. At just 25 minutes before ten o'clock and just 7 minutes after the stage with its load of passengers crossed the bridge, that structure was carried away. The huge mass of driftwood struck the middle pier which supported the bridge and carried it away. In less than a minute, thereafter, the bridge was afloat, and a minute or less afterward, it broke in two pieces and was carried down the river with other driftwood, logs, trees etc. When the driftwood struck, the center pier was carried away and at the same moment the end of the bridge at the Yuba County side swung around and in a moment thereafter, the whole bridge was afloat.

At Smartsville the column of water was five feet above the normal level and was a solid wave, bearing on its crest a compact mass of logs and other driftwood. It formed what appeared to be a floating bridge, sweeping everything before it and threatening 40 head of cattle. Fortunately, none were drowned. Pine trees, two feet thick were still being uprooted by the wall of water flowing at the rate of ten miles an hour. They were carried along until actually stripped of their bark and completely splintered.

Upon reaching Marysville at 3:10 P.M., the water was two feet above the normal level. The levee at Linda, just above the city, was broken by the force of the flood, spreading its turbulent waters over the surrounding land. This weak spot in the restraining wall was considered a boon for the City of Marysville.

The English Dam located at the South Fork of the Middle Yuba River, about 42 miles northwest of Nevada City, was actually formed by three dams. The walls were constructed of dry rubble stone, covering a solidly filled timber crib. The central section was 131 feet high and 400 feet long. The reservoir fed 80 miles of ditches with a carrying capacity of 2,800 inches, leading to the Milton Company's hydraulic mines at Badger Hill, Manzanita Hill,

Birchville and French Corral. First constructed by the English Company in 1859 at a cost of about $75,000, the lake was two and one-half miles long, and one-mile wide, covering 395 acres and contained 618,000,000 cubic feet of water. The principal owners at the time of the disaster were: Thomas Bell, L. L. Robinson, Egbert Judson, Hamilton Smith, Jr. and the estate of V. G. Bell. The North Bloomfield Gravel Mining Company held the controlling interest.

The dam disaster caused about 100 hydraulic miners to be thrown out of work — 50 at French Corral and possibly another 50 at Sweetland and Sebastopal. The Eureka Lake Company's flume which crossed Bloody Canyon was carried away and damage to the American Mine at Sweetland was thought to be in the range of $5,000. A temporary ford was constructed at Freeman's Crossing, allowing the stage with a team of eight horses to cross safely. Freeman gave assurances that he would build a new bridge with all haste and within a week he had 50 men working to replace the bridge and roads. The loss to his property was estimated to be in the neighborhood of $12,000. His covered bridge at Oregon Creek was found lodged 150 feet below its original location. Its stone abutments were rebuilt and the bridge was hauled back over log rollers by a team of oxen and lifted in place—except that it was end-for-ended. The south end was now facing north. This sturdy, 80-foot structure is still in use today, spanning Oregon Creek.

The *Appeal* estimated the damage to Marysville as follows:

Standing grain	$25,000
Grain in stack	5,850
Pasturage	1,040
Potatoes, etc.	2,000
Repairing Levee	<u>1,000</u>
Total	$35,450 [$34,890]

H. C. Perkins, superintendent of the North Bloomfield and the Milton companies, revealed there were strong indications that the dam had been blown up by powder, and immediately posted a cash reward of $5,000 for the arrest of whoever was responsible. Perkins also placed a watchman at Bowman Dam to prevent the possibility of a similar disaster. As the *Transcript* dramatically put it ". . . the watchman is Bob Hamilton, a man who knows no word as fear and who is an unerring shot with the rifle and pistol. From dusk to dawn he maintains his lonely vigil in the solitude's of this mountain fastness, keeping a vigilant look-out for suspicious or unknown characters and warning them off when they haunt forbidden localities."

On the other hand, it was the contention of the valley residents that the dam failed because it was overtopped. This was an assertion Perkins vehemently denied, stating that it had been a rainless month and the dam had had a critical inspection only three days before the break.[43]

When the English Dam was transferred to the Milton Company in 1870, it was in need of some serious maintenance. Even as late as the time of the heavy snows of the winter of 1875-76, the condition of the dam was a matter of great concern to Hamilton Smith who was thinking of the heavy runoff that would come in the following June and July. Writing to Superintendent V. G. Bell of the Milton Mining & Water Company at French Corral, on March 31, 1876, Smith cautioned ". . . the middle and north dams of this reservoir are in such bad condition that the trustees are of the opinion that it will be unsafe to allow the water, during the coming season, to raise higher in the reservoir than the bottom of the present waste-way." He then requested Bell to have a new opening cut lower than the present waste-gate. On June 15, in another letter addressed to Bell, Hamilton Smith mentioned that he was preparing to take a trip to English Dam "to determine if repairs are necessary this year." The following July 4th, he again wrote Bell from Bowman Dam, discussing the work that needed to be done which, among other things, included removing rocks from the top of English Dam. He explained in some detail exactly how he wanted it done. Smith also suggested to Bell that he use Chinese labor, "which was used at Bowman Dam" and referred to as "very fair quality of Chinese labor." He also remarked that he paid them $1.15 per day in wages.[44] Later the Milton Company began facing the dams with a massive stone wall which at the time of the accident, was completed to within six feet of the water level. The remainder of the wall, at that time, was a temporary wooden structure.

Whatever the cause of the English Dam disaster, both sides agreed, it hastened the Sawyer Decision. Also, in a sense, it served as a dramatic and tragic precursor to the conclusion of thirty years of hydraulic mining in the Sierra Foothills.

On January 7, 1884, Judge Sawyer's decision was made known—the miners were enjoined perpetually and unconditionally.

> Exactly at one o'clock [the *Appeal* related] the first message was received [at the telegraph office] announcing in the most positive terms a complete victory for the valley. In a few moments the steam whistle at Swain & Hudsen's mill began a shrill prolonged shriek, which soon appraised everyone throughout the city that the glad tidings had come at last. Other steam whistles took up the burden of the story, and soon the church bells were ringing out the most joyful peals ever heard in Marysville. In a few minutes the streets were filled with people shaking each other by the hand, cheering and voicing the satisfaction and delight in the most hearty terms. Flags began to appear over buildings, saloons became jammed. The decision occupies 400 pages. An anvil salute was fired at the Empire Foundry. At five o'clock a salute of 40 guns was fired from Captain Colford's battery on the levee at third and G Streets. A huge bonfire was started at Cortez Square fed by boxes, crates, barrels and casks 20 feet in height, crates and boxes filled with straw, lit early in the evening. Hundreds of people gathered cheering to the sounds of a brass band. Later that night the crowd was treated to a brilliant display of fireworks from the roof of the *Appeal* office. All of the fireworks found in the city were consumed.[45]

In Nevada City the mood was quite different. "It would be difficult to find in all the law reports a more partisan decision than that Judge Sawyer recently delivered in the debris case of Woodruff vs. the North Bloomfield Mining Company *et al*," Calkins, editor of the *Transcript* snarled. "Since the inception of this case in this Court until the close of the delivery of this most unjust and partisan decision, his prejudices against the miners has been so prominent in all ways that it has been the subject of frequent comment." Calkins no doubt hoped his former accolades to Judge Sawyer had long been forgotten.[46]

The Oakland *Times* probably summed it up best. "It was expected by both the farmers and hydraulic miners, that the slickens case would result in a compromise decision, and Judge Sawyer's finding therefore produces on one side great depression and on the other great exultation."

But as a landmark case in the legal history of the state, the *Appeal*, no doubt, expressed the views of most Californians. "It blots out a relic of barbarism and sustains civilization. It declares and reaffirms that great legal maxim which is the foundation of all law and civilized society—that no man has the right to destroy his neighbor's property because it is profitable to the trespasser." [47]

∗ There have been various estimates of the amount of material that was washed from the Sierra into the San Joaquin and Sacramento rivers and their tributaries and into the great valley and, eventually, into the San Francisco Bay. Lindgren's appraisal concluded that it amounted to 1,295,000,000 cubic yards of debris. Other estimates ran as high as 2 billion cubic yards of slickens that were washed from the mountains. Bowie calculated that 15,122,000 miner's inches of water were used in hydraulic mining annually. From this figure he concluded that each year 53,464,000 cubic yards of material were removed. Another popular estimate was 46,000,000 cubic yards annually—an amount which would form a mass a mile wide and a mile long and 15 yards deep. In another study, using 1880 as a base year, it was determined there were more than 400 hydraulic mines reported to be in operation within the drainage areas of the Sacramento and San Joaquin rivers. During that year over 10,000,000 inches of water were used, leading to the conclusion that 38,000,000 cubic yards of gravel were mined in that year. [48]

∗ G. K. Gilbert the celebrated geologist, then with the United States Geological Survey, at the specific request of the Miner's Association, was commissioned by President Theodore Roosevelt to make a determination of the amount of mining debris that had actu-

Manzanita hydraulic mine, near Sweetland, Nevada County, owned by the Milton Mining & Water Co. Photo taken by G. K. Gilbert of the U. S. Geodetic Survey. Gilbert made measurements of hydraulic cavities to determine amount of debris washed from the Sierra.

ally been washed into the San Francisco Bay. It was Gilbert's belief that the methods used to make previous calculations were faulty. They were based on a computation of the amount of water used in miner's inches and then assigned a duty to the inch. Gilbert used an entirely different method which was based on the careful measurement of the cubic yards extracted by the excavations. With these calculations, he determined that over 1.6 billion cubic yards had been washed into the Bay and that it would take about 50 years before the mountain streams would be free of debris.[49]

The Sawyer Decision effectively shut down or severely crippled the larger hydraulic mines, but a number of the smaller ones persisted in their operations many with a lease arrangement. The Ant-Debris Association doggedly maintained organized opposition, resulting in long and costly litigation, continuing for many years.

Since an injunction was required for each violation, the Anti-Debris Association constantly kept men in the field to secure evidence. Feelings, of course, ran high against these "spies," as they were derisively dubbed by the miners. A tip from one, led to the deployment of a gang of deputy sheriffs from Marysville to Emigrant Gap on a nighttime trip by special train. From there they hiked into the Omega Mine and at daylight surrounded a bunkhouse and seized 20 Chinese hydraulickers. After being taken to Marysville, the Chinese were given their choice, by a judge, of paying a fine of $500 each or working on the county roads for 500 days. All twenty agreed to work on the roads, but after the judge learned what it was costing the county to maintain the prisoners, all 20 were discharged. The Omega Mine, however, was found in contempt of court and fined.[50]

It should be noted that even before the Sawyer Decision, the attitude in the foothills toward Chinese hydraulic miners was undergoing a change. As the Auburn *Herald* reported ". . . Very few [hydraulic miners] are going to any considerable expense in fitting up in consequence of the uncertainty of being allowed to continue. The slickens agitation . . . is proving a grand harvest for the Chinamen. Nobody shows any disposition to molest them, and as a result they are either buying or leasing nearly every mine that has anything in it."[51] Indeed the Chinese population in the mines was actually profiting from the anti-debris legislation. From the pages of the Howland Flat *Sierra County Tribune*, it was reported that at Scales Diggings, a prominent mining town: "The outlook for white men to exist in these parts is getting very slim. Chinese labor is becoming more popular every day among mine owners." The following month the same newspaper disclosed that at Howland Flat itself, "The center of town occupied by Chinese, presents a lively and business-like appearance. The part of town occupied by the white population is extremely dull."[52]

Many of the hydraulic mines attempted drifting the gravel, some with moderate success. Hirschman & Company near the southwest border of Nevada City, for a while, found it quite profitable. Other reports indicated that the old Knickerbocker hydraulic workings

THE ENDING 259

were drifting "very fine gravel." At Lowell Hill and Remington Hill some remunerative gravel had been found and a new tunnel had been run. But at most hydraulic claims, the ground was just not rich enough to drift profitably. A report from French Corral indicated that miners ". . . were scraping up the gravel from the bedrock and wheeling it to their sluices where the washing is done. It will take only a few weeks to work it all out, and then they will be compelled to seek other places to labor." A news item from Yuba County was becoming quite familiar, ". . . all the hydraulic mines in Smartsville shut down and it is felt there is little hope of reopening. Many of the unmarried men have moved away."[53] Quoting the Marysville *Appeal* the *Sierra County Tribune* reported: "President Sexey of the Anti-Debris Association, has received a communication from Supt. Lawrence of the Brandy City Hydraulic Mining Company in which he states that the company has closed operations and is disposed to be law abiding . . . The mine is represented as being too shallow to drift, and it is supposed that as it cannot be worked under the hydraulic process it must be abandoned."[54]

In reply to a report by the *Transcript* that the North Bloomfield Company was hunting for gravel that would pay to drift, the Sacramento *Bee* exclaimed: "Yes hunting for it with streams of water from little giants, we suspect. Men armed with rifles guard the North Bloomfield ground from intruders. Monitors have been heard working the mine at night. The company has been enjoined by a federal court, but there is reason to believe that it is systematically violating Judge Sawyer's decree. . . . Means must be found to check the criminal proceedings of such sneaking scoundrels."[55]

But according to August J. Bowie, ". . . the decision of the court applied to all classes or kinds of mining, there can be no doubt." Suits had been instituted, for example, against parties for drift mining ". . . and in one instance the North Bloomfield Gravel Mining Company has been adjudged guilty of contempt of court, and fined heavily for violating this injunction by drift mining."[56]

Hydraulic mining was not declared unlawful by the courts, but only as it had been previously conducted, where it contributed or threatened to contribute in a material degree to the filling up of river channels, injuring navigation, or covering the lands adjacent to the navigable streams with debris. Where these conditions prevailed, such actions were declared a public nuisance and prohibited. Judge Sawyer did make certain suggestions of possible conditions under which the injunction might be dissolved. It will be noted that his decree was not against hydraulic mining, but dumping the debris into the streams as outlined above. Therefore, it could be and was applied to other forms of mining.

At the time of the Sawyer Decision there was estimated to be $100,000,000 invested in hydraulic mining in the state. A report from the Assembly Committee on Mines placed this figure even higher stating: "There has been expended in cash, aside from the intrinsic value of the mines, in building canals, reservoirs and tunnels, for the purpose of working these gravel channels, a sum of not less than one hundred and thirty million dollars up to the present date [1876]."[57] During the ensuing period of long and costly litigation and bitter controversy, mine after mine was closed down and the mining property, ditch systems and reservoirs became worthless. The water companies which sold the miners water were enjoined from such sale and their properties also became worthless. The great water systems of ditches, flumes, reservoirs and miles of pipeline, in many cases, were simply abandoned. Dams were even blown up and destroyed. Throughout the hydraulic mining regions, thousands of men were thrown out of employment with camps and whole mining towns deserted. During the ten years from 1880 to 1890, the assessed valuation of property in Nevada County declined by $10,000,000 and with the dwindling population the rate of taxation doubled. A period of ten to twelve years would pass before these regions would again gain their equilibrium.[58]

Although Nevada County had the largest concentration of hydraulic mining operations in the state and initially suffered the most from their closures, it likewise, possessed the greatest diversity of gold-bearing deposits found in California. The county, thus in a few years, displayed a remarkable resilience. Even on the nearly abandoned San Juan Ridge, unsurpassed in the quantity of its auriferous gravel deposits, new quartz veins were being discovered. Near Columbia Hill, the Delhi mine was attracting wide attention from the richness of its vein. Other recent discoveries on the divide included the Golden Gate, Delaware, Boss, Tip-Top, Little Giant and Golden Wonder to name but a few. In the Washington District, older lode mines that had long been idle or had been poorly developed were being reopened and outfitted with new mills, and were becoming conspicuous as producers of bullion. Among these mines were the Yuba, the Eagle Bird, the Cornucopia and the Secret Treasure.[59]

Placer County, on the other hand, lacked the versatility of its neighbor to the north and, thus, suffered more. Some of its quartz mines were actually on the decline, as well, such as those in the Penryn, Ophir and Auburn districts. But the damage caused by the closing of its hydraulic mines, according to the *Mining and Scientific Press*, was incalculable.

> The misery [is] appalling. Flourishing towns have in consequence been depopulated. Men, well-to-do only a few years ago, have been financially ruined, and even whole communities brought to the verge of bankruptcy. Hundreds of industrious and well-paid men have been converted into tramps—forced to leave their homes and seek employment elsewhere, it being impossible for them to get work in the mines. Families have been broken up and scattered, the fathers being no longer able to keep them together. Other industries throughout the enjoined districts, suffering through the destruction of the main stay, have languished and died; and so desolation broods over the homes, and silence reigns throughout a country once enlivened by the roar of a mighty industry.

On the Forest Hill Divide from Last Chance to Todd's Valley, a distance of 20 miles, hydraulic mining everywhere was dead, causing a scarcity of money and a general depression in all business. At some of the hydraulic mines drifting was tried but in most cases, with such poor results, it had been given up.

In the Southern Mines or Mother Lode, where hydraulic mining was a minor source of bullion production, the effect of the Sawyer Decision was much less severe. In Calaveras County at Chili Gulch, the Duryea mine, for years a profitable hydraulic mine, had successfully switched to drifting and became one of the best paying operations in the county.[60]

While it is impossible to give precise figures with respect to the total gold production from hydraulic mining up until 1884, because official reports did not segregate the various forms of placer mining, it is significant that placer gold production from the period, 1871 to 1880 was $121,000,000, while that of the following decade, 1881 to 1890 was $38,000,000 and the following decade, 1891 to 1900 was a mere $29,000,000. Hydraulic mining produced most of California's gold between the late 1860s and 1884. After the Sawyer Decision, production, for all methods of mining, hit its lowest point around the year 1887, then with improved lode-mining techniques and the advent of dredging, production continued to increase. From 1871 to 1880, 70 percent of all gold produced was by the placer method, but in the following decade that percentage would drop to 25. According to Arthur Jarman, the highest annual production from hydraulic mining was $15,000,000, with a conservative annual average of no less than $10,000,000 for the 30 year period from 1853 to the Sawyer Decision—making a total production of $300,000,000. Most authorities agree with this figure.[61]

During the early months of 1891 a number of Placer County miners, who were particularly hard pressed as a result of the closing of the mines, agreed to get together and see if something could be done to get them open again. At this informal meeting, it was decided that a convention of Placer County miners should be called to discuss the matter. From this small beginning, a statewide convention was subsequently held in San Francisco with all the mining counties in the state represented. Even farmers from the valley were in attendance. After mutual concessions an amicable common plan was agreed to, which was based upon an earlier report of a government commission of engineers. At this point the matter was placed before the State Legislature, which passed a joint resolution, bringing the matter to the attention of the United States Congress. Without too much hesitation, a commission was established to follow through with proper recommendations.

As a direct result of these initiatives, in March 1893, Congress passed the so-called Caminetti Act which permitted gravel mines to be operated by the hydraulic process with certain restrictions and conditions. Under the Act, the California Debris Commission, consisting of three officers of the Corps of Engineers, was appointed by the president. The Commission was empowered to issue licenses for mining by the hydraulic process, when it was satisfied that proper restraining dams had been constructed to impound the tailings. A tax of three percent of gross revenues from hydraulic mining would be set aside in a "debris fund." This would be administered by the Commission and used for the purpose of erecting debris dams, together with any other moneys from government sources.[62] To facilitate the application of the Caminetti Act, the legal status of hydraulic mining was spelled out in the State Legislature in two sections of the Civil Code of California:

Section 1424. The business of hydraulic mining may be carried on within the State of California wherever and whenever the same can be carried on without material injury to the navigable streams or the lands adjacent thereto.

Section 1425. Hydraulic mining within the meaning of this title, is mining by means of the application of water, under pressure, through a nozzle, against a natural bank.

The California Debris Commission Act was passed, not only to protect farming lands and navigable rivers, but was also intended to assist hydraulic mining. The miners expected it to accomplish great things for them. It provided a nonpolitical body of army engineers whose duty it was to administer the act, survey the rivers for dam sites, and erect dams when money was appropriated for them. But the dams were never built and hydraulic mining never experienced a significant revival. Another reason often cited for the failure of a resuscitation of the ailing industry was the cost and relative ineffectiveness of the restraining dams. The hydraulic miner first had to apply to the Commission for a license by submitting plans showing the proposed restraining works. This was followed by a personal examination of the site by a commissioner. Each application then needed to be advertised for a specified period of time. Then a hearing was held before the Commission where anyone opposed to the issuance of a license could speak his piece. If the plans were approved and the dam constructed, it would then be inspected at intervals to make sure no debris was entering the mountain streams.

Many of these debris dams, even after they were approved, often failed following a period of heavy rains. An April 1895 issue of the *Transcript* described an impressive new dam to be constructed on Scotchman Creek in Washington Township. It was to be 250 feet long and 65 feet high with a base 120 feet thick. The following November, in a follow-up story it was learned: "The restraining dam built in the Omega Hydraulic Mine by the Tully Bros., is pronounced by the Debris Commission to be the best dam for the purpose in the state. John Wart, the photographer, went to Omega yesterday to take views of the dam for the Commission." But the following January it was learned that at the Omega Dam, during a heavy storm, after days of heavy rain, the spillway shaft became clogged and the dam gave way. The story added that three Chinese miners were swept away and only one body, that of Ah Get, was recovered.[63] But not all restraining dams were destroyed by storms. In a number of documented cases the deed was accomplished by arson or blasting. Eight years after the disaster at Scotchman Creek another operator at the same mine, Colonel Joseph Underwood, constructed a log and rock dam of even more impressive dimensions, 79 feet high and 1138 feet long. On January 2, 1905, the *Transcript* related a not too unfamiliar tale. "Friday night the new debris dam, just completed by Colonel Underwood's company, in Scotchman Creek was carried away. . . . Everything indicated that dynamite was used to destroy the dam."[64]

Hydraulic mining was never again a viable method in the Sierra foothills. According to Waldemar Lindgren, from 1897 to 1909, only $5,858,353 was produced in the mining counties north of Mariposa in the Sierra Nevada range by this method. This total was

THE ENDING

Debris-restraining dam below the Red Dog Hydraulic Mine, Nevada County.

calculated to be an average of just $330,136 per year. Nevada County produced the most with a gross of $1,509,986. Calaveras County, the greatest producer for counties draining into the San Joaquin, totaled $389,200 for the same period. In 1925 the production from hydraulic mining in California was a mere $175,345 and, of that amount, over half came from areas exempt from any restrictions.[65]

But, according to the mining industry and government geologists, there was still an abundance of Tertiary gold in the Sierra foothills. Although some of these reports revealed a degree of variation, they all agreed that the gravels that remained were extremely rich. The most optimistic report, noting the potential for hydraulic mining, was made years earlier by Rossiter W. Raymond in his *Mineral Resources of the United States*, where he estimated that, in California, from

$1,000,000,000 to $1,500,000,000 could be produced by hydraulic mining.[66] Arthur Jarman noted that the yields of the Yuba, Bear and American River drainage areas had values of 11.5 cents per cubic yard for over 200,000,000 cubic yards of gravel that still remained. Other reports indicated that on the San Juan Ridge alone, 408,000,000 cubic yards remained at 10.3 cents with yet another 25,000,000 cubic yards valued at 28 cents. On the American River 40,000,000 cubic yards of auriferous gravel remained with values of 18 cents per cubic yard. Boyle estimated that in Nevada County alone, between $75,000,000 and $100,000,000 in gold ($35.00 per ounce) lies buried in the gravels of the Neocene streams between North Bloomfield and French Corral.[67] In a more recent report using modern technology, including aerial photography and extensive core drilling, the area between the Middle Yuba River and the North Fork of the American River was thoroughly explored. As a result of a portion of this study, it was determined that the San Juan Ridge, between Malakoff State Park and Badger Hill, contains 800,000,000 cubic yards of gravel worth $140,000,000 at $35.00 prices.[68]

During the depression years of the early 1930s, there grew an increasing desire to, once again, tap this great wealth. It was reasoned that the greatest obstacle to hydraulic mining, following the Caminetti Act, was the lack of high debris dams. Under Section 23 of that act, authorization was given for their construction, but no funds had been granted. The cost was to be provided by storage charges—a very unrealistic notion. In 1934 Representative Harry L. Englebright of Amador County succeeded in having an amendment to the Caminetti Act passed that provided for a more feasible method of repayment. The following year, the Rivers and Harbors Committee in the Senate was successful in obtaining an appropriation of $6,945,000 for the construction of four high dams. Four damsites were proposed; one on the Upper Narrows on the Main Yuba; another at Dog Bar Site on the Bear River; the North Fork Site on the North Fork of the American River; and finally, the Lower Ruck-a-Chucky Site on the Middle Fork of the American River.

THE ENDING

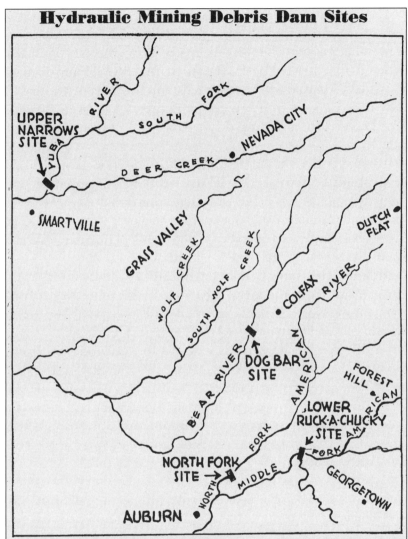

The sketch above shows the approximate locations of the debris dams, approved and awaiting approval by the United States War Department as the first step in the resumption of large scale hydraulic mining in Sierra, Nevada and Placer Counties. In the lower right portion, near Auburn, are the sites for the dam to be built on the North Fork of the American River and the one at Ruck-A-Chucky on the Middle Fork of the American. In right center, on the Bear River, is the site of the proposed Dog Bar dam. In the upper left, above Smartville, is the location of the proposed Upper Narrows dam, to be erected on the South Fork of the Yuba River. The guarantees of repayment of costs to the government have not yet been presented to the army engineers.

The long years of depression had produced a great change in attitude on the part of the Valley people with respect to the resumption of hydraulic mining. The Sacramento *Union*, in a mood of optimism declared that it would be a boon to the city. "Approximately $15,000,000 in gold will be recovered and a new river of wealth will flow into Sacramento if congressional approval is given the proposed construction of three debris dams in the mountains east of here."[69]

Inspired by the euphoria following the completion of the dams, a new organization named the Hydraulic Miners Association was formed complete with its own publication titled *The Hydraulic Miner*, and claiming 1,000 dues-paying members. But after a brief two years it suspended operations "due to lack of finances."[70] Once again, great expectations came to naught.

Although the Harry L. Englebright Dam was constructed at the Upper Narrows of the main Yuba and, another, the North Fork Dam, on the North Fork of the American River, and although the price of gold had been increased in 1934 from $20.67 to $35.00 per ounce, the measure was a classic case of—"too little too late." It was too little, because for any hope for success, four high restraining dams were necessary, and too late, because the vast quantities of water required for successful hydraulic operations had long ago been dedicated for purposes of irrigation and for hydroelectric power. Less than three percent of the storage capacity of the two dams was ever utilized.[71]

All hope for the revival of hydraulic mining in California was over—forever!

Appendixes

APPENDIX A

Gold Production in California

Gold Production for the First Twelve Years in California Based on Customs House Receipts at San Francisco

Year	Amount
1849	$4,921,250
1850	27,676,346
1851	42,582,346
1852	46,582,695
1853	57,331,034
1854	51,328,653
1855	45,182,631
1856	48,887,543
1857	48,976,697
1858	47,548,025
1859	47,640,462
1860	42,303,245

An accurate determination of the gold production in California is extremely difficult. The best basis is to take the deposits of domestic gold made at the mint and the exports of uncoined bullion. However, the statistics for both of these sources are defective before the year 1855. Therefore estimating is required. Another problem is the determination of how much of the gold was used for private coinage made in California by the old United States Assay Office and other coiners from 1849 to 1855. It is estimated that at least $60,000,000 was coined in those seven years. Much of it was exported as soon as it was made. It is believed that there was no less than $25,000,000 to $30,000,000 in circulation when the mint went into operation in April 1854.

Also in the early days large amounts of gold were taken out of the state in private hands, because of the high rate charged by the steamers for the export of gold in any form. The result was that for several years the deposits at the eastern mints exceeded by 10 to 15 million dollars annually the entire export from San Francisco as shown by the customhouse records.

Every steamer carried from 500 to 1000 passengers. In addition, between 1862 to 1872 considerable gold was taken in this manner by returning Chinese.

Standard Estimates of Gold Production

Year	Amount
1848	$ 245,301
1849	10,151,360
1850	41,273,106
1851	75,938,232
1852	81,294,700
1853	67,613,487
1854	69,433,931
1855	55,485,395
1856	57,509,411
1857	43,628,172
1858	46,591,140
1859	45,846,599
1860	44,095,163
1861	41,884,995
1862	38,854,668
1863	23,501,736
1864	24,071,423
1865	17,930,858
1866	17,123,867
1867	18,265,452
1868	17,555,867
1869	18,229,044
1870	17,458,133
1871	17,477,885
1872	15,482,194
1873	15,019,210
1874	17,264,836

APPENDIX A

Total Gold Production in California 1848 to 1926

$1,801,106,027

Estimate by James M. Hill, "Historical Summary of Gold, Silver, Copper, Lead and Zinc Produced in California, 1848 to 1926," U.S. Bureau of Mines, *Economic Paper*, No. 3 (Washington:1929).

Total Gold Production in California 1848 to 1968

$2,428,330,901

Estimate by William B. Clark, "Gold Districts of California," California Division of Mines and Geology, *Bulletin* 193 (Sacramento: 1976). This estimate is based on the price of gold at $20.67 until 1933 when it was raised to $25.56. It then rose to $35.00 in 1935. On March 15, 1968, gold was sold on the open market.

Total Gold Production for the Six Counties Comprising the Southern Mines (Mother Lode)

$724,000,000

Total Gold Production for the Six Counties Comprising the Northern Mines (north of the Cosumnes River)

$1,110,000,000

APPENDIX B

Ratio of Placer Production to Lode Production

Period		Percent
1848-1850		100%
1851-1860		99
1861-1870		90
1871-1880		70
1881-1890	(Sawyer Decsion)	25
1891-1900		20
1901-1910	(Beginning of Dredging)	35
1911-1920		44
1921-1930		44

From J. H. Hill, *Economic Paper* No. 3

The 50 Leading Hydraulic Mines in California by County

County	Number
Nevada	18
Sierra	13
Placer	11
El Dorado	2
Plumas	2
Amador	1
Butte	1
Trinity	1
Yuba	1

From William B. Clark, *Bulletin* 193

APPENDIX B

Principal Hydraulic Mines

Mine	Location
Alpha	Washington district, Nevada County
Badger Hill	Badger Hill district, Nevada County
Brandy City	Brandy City district, Sierra County
Buckeye Hill	Scotts Flat district, Nevada County
Cherokee	Cherokee district, Butte County
Cherokee	Badger Hill district, Nevada County
Chips Flat	Alleghany district, Sierra County
Craigs Flat	Eureka district, Sierra County
Deadwood	Last Chance district, Placer County
Depot Hill	Indian Hill district, Sierra County
Dutch Flat	Dutch Flat district, Placer County
Elephant	Volcano district, Amador County
French Corral	French Corral district, Nevada County
Gibsonville	Gibsonville district, Sierra County
Howland Flat	Poker Flat district, Sierra County
Indian Diggings	Indian Diggings district, Eldorado County
Indian Hill	Indian Hill district, Sierra County
Iowa Hill	Iowa Hill district, Placer County
La Grange	Weaverville district, Trinity County
La Porte	La Porte district, Plumas County
Last Chance	Last Chance district, Placer County
Liberty Hill	Lowell Hill district, Nevada County
Lost Camp	Emigrant Gap district, Placer County
Lowell Hill	Lowell Hill district, Nevada County
Malakoff	North Bloomfield district, Nevada County
Mayflower	Forest Hill district, Placer County
Michigan Bluff	Michigan Bluff district, Placer County
Minnesota	Allaghany district, Sierra County
Moore's Flat	Moore's Flat district, Nevada County
Morristown	Eureka district, Sierra County
North Columbia	North Columbia district, Nevada County
Omega	Washington district, Nevada County
Paragon	Forest Hills district, Placer County
Port Wine	Port Wine district, Sierra County
Poverty Hill	Poverty Hill district, Sierra County
Red Dog	You Bet district, Nevada County

Mine	Location
Relief	North Bloomfield district, Nevada County
Remington Hill	Lowell Hill district, Nevada County
Sawpit Flat	Sawpit Flat district, Plumas County
Scales	Poverty Flat district, Sierra County
Scott's Flat	Scott's Flat district, Nevada County
Smartsville	Smartsville district, Yuba County
Stewart	Gold Hill district, Yuba County
Texas Hill	Placerville district, El Dorado County
Todd Valley	Forest Hill district, Placer County
Whiskey Diggins	Gibsonville district, Sierra County
Yankee Jim's	Forest Hill district, Placer County
You Bet	You Bet district, Nevada County

APPENDIX C

Storage capacity of hydraulic mining reservoirs in California, 1883 — 7,600,000,000 cubic feet

Mine/Ditch Company	Storage Capacity (cu. ft.)
South Yuba Canal Company	1,800,000,000
Eureka Lake & Middle Yuba	1,130,000,000
North Bloomfield Company	1,130,000,000
El Dorado Deep Gravel Co.	1,070,000,000
Milton Mining & Water Co.	650,000,000
California Water Company	600,000,000
Spring Valley Mining Co.	300,000,000
Omega & Blue Tent	300,000,000

Compiled by Taliesin Evans, *Hydraulic Gold-Mining in California*

Principal Ditches in California Gold Fields

North Bloomfield Co.	Excelsior or China Ditch
Milton Mining Co.	Bouyer and Union
Eureka Lake Co.	El Dorado
San Juan	Cherokee/Spring Valley
South Yuba Canal Co.	Hendricks
La Grange	

Compiled by Turrentine W. Jackson, "Report on the Malakoff Mine"

Distribution of Auriferous Gravel Remaining in the Sierra Nevada — 7,000,000,000 Cubic Yards

District	Amount (in cubic yards)
Feather River	500,000,000
Yuba River	3,500,000,000
Bear/American River	2,500,000,000
San Joaquin Watershed	500,000,000

Compiled by Charles Scott Haley, "Gold Placers of California"

APPENDIX D

"Hydraulic Mining," (as some believe), described by Pliny as it was practiced in Spain. More properly it should be considered ground sluicing. Pliny referred to it as "arrugea."

In carrying out the "arrugea," the mountain selected was first undermined by the labor of innumerable men [slaves]. Their means of penetrating the rock consisted in hammers, drills, grads or wedges, and now and then the pick-axes of huge size. These they assisted with fire, heating the rock, in order to soften it; and the effects of the fire were enhanced by the use of water and vinegar — the latter being especially efficient upon limestone. Under these circumstances, of course their progress was slow, and many months were consumed in excavating the mountain and removing all the loosened material. During this work the premature fall of the overlying mass was prevented by means of rock pillars, left standing in the mine. Finally these pillars also were hewn away, at much risk of life, and the mountain fell with a great crash. A sentinel, placed on the outside, observed the first appearance of fissures, and signaled this fact to the miners underground — after which —*sauce qui peu.*

These dangerous and costly operations were followed by others not less complicated and wonderful, having for their object the separation of gold ores from the *debris* of rock. Brooks and rivers were conducted, sometimes from great distances, to the locality, that their united volume might be employed in washing away the earth and the waste rock. No difficulties were too great for these ancient engineers — Valleys were crossed with aqueducts; hills were pierced with tunnels; the waters thus brought together were at last collected in tanks, two hundred feet wide and ten feet deep, and from these precipitated upon the broken masses of the mountain with such force that even large boulders were moved by the stream.

APPENDIX D

But since so powerful a current would unavoidably carry away also many particles of gold, the water was again collected below in many wooden channels, the bottoms of which were covered with a certain plant called *Ulex*, resembling our rosemary. This rough plant answered the purpose now served by coarse blankets in the sluices of California. It caught and held the particles of gold and these were finally obtained by drying and burning the herb.

American Journal of Mining, Vol III, No. 14

APPENDIX E

Initial letter from Hamilton Smith, Jr. Leading to the formation of the Hydraulic Miner's Association

SanFrancisco, July, 1876

DearSir,

It has been deemed desirable by several gentlemen of this City; who are largely interested in hydraulic gravel mining in California, that an association should be instituted composed of members residing in various parts of the State, who are owners either of gravel mines, or of canals for the supply of water to such mines, or of tail sluices:theobjects of the proposed association to be the interchange of useful knowledge between the members in everything relating to the art or process of gravel mining and also for general purposes of mutual benefit and protection.

A meeting of some of the gentlemen was held in San Francisco on April 26, 1876, when initiatory steps were taken towards the formation of such an association.

It was there proposed to effect an organization in the following described manner:

The office of the association to be in San Francisco, and to be a place of general resort for its members when they chance to be in the City.

The government of the association to be vested in a Board composed of 17 Trustees, five of whom shall be residents of San Francisco and the remaining 12 members to be selected from different counties in the State, where the business of gravel mining is extensively carried on.

The five resident Trustees to constitute a quorum among themselves for the management of ordinary business, provided none of the non-resident Trustees are present, but who, if present shall fully participate and vote in any meeting of the Board. If, however, any extraordinary action should be necessary, such as a change in the By-Laws, then a majority of the entire Board will be required to constitute a quorum. These Trustees to have absolute power to direct the affairs of the association, and to carry out the business for which it was organized.

The Trustees to elect from their own number a President, who shall receive no salary, unless his services for any considerable length of time be especially devoted to the interests of the association; the Trustees to decide as to the necessity, if any, of such employment, and the rate of compensation.

A Secretary to be elected by the Board, who will be expected to remain during office hours in the office of the association, and to keep the accounts and conduct its correspondence.

The rent of the office, stationery and the salary of the Secretary are expected to be the only costs of general management.

The expenses of the association will be defrayed by a small initiation fee from each member, and assessments or dues levied upon the mining and ditch properties of its members as assessed for state taxation. It is expected that this assessed value of the properties of the members will amount to several millions of dollars, and that hence the cost to each member for his share in supporting the association will be but trifling.

The Trustees will be elected by an annual ballot of the members, and each member will have a number of votes in direct proportion to the taxable value of property he represents: that is to say, he will cast more or less votes, just as he pays more or less dues.

Each member of the association will be solicited to furnish to the Secretary all available and interesting information within his reach as to new and improved methods of mining, tunneling, ditching, &c.,&c., and also general statistical facts, as to the cost and yield of various mines when the owners are willing to publish such information: the Secretary will disseminate among the members, carefully collated, all this knowledge, and in this and other kindred ways the association will prove to be of great practical benefit.

It is also proposed that the association shall take upon itself, whenever in the opinion of the Trustees it may be deduced just and advisable, the defense and protection of any one of its members who may be legally attacked, and when the point at issue involves some general principle, in which the other members are also interested. In this way the costly and vexatious expenses of long contested litigation, will be defrayed, not by a single individual or corporation, to whom such legal costs might be absolutely ruinous, but by the whole community of gravel miners, on whose united shoulders the burden will fall, and where it properly belongs.

In any such legal contest, where the defeat of the owner of a claim 40 feet square might most seriously imperil the value of all our gravel mines and water properties, it is most apparent that we should defend our just rights in a united body, and not allow the destruction of a single weak member to result in the destruction of us all.

As the costs of the operations of the association will be defrayed by those chiefly concerned in its management, it is hoped and believed that its affairs will be conducted in a most economical and also judicious manner.

It is proposed to hold the first Annual Election on Monday, September 4th, 1876, at the office of the North Bloomfield Gravel Mining Co., No. 320 Sansome St., in this City; where the full Board of Directors will be elected and the permanent organization finally completed.

You are invited to attend said meeting to represent your property, and protect its interests by assisting in effecting this organization.

The following named gentlemen of San Francisco have been designated as temporary Trustees to hold office until the first annual meeting is held.

J. P. PIERCE, representing Excelsior Water Co. and California Water Co.

EGBERT JUDSON, representing Spring Valley Mining and Irrigating Co. and Milton Mining and Water Co.

L. L. ROBINSON, representing Union Gravel Mining Co.

THOMAS PRICE. Representing El Dorado Water and Deep Gravel Mining Co. and Blue Tent Consolidated Hy. M. Co., limited.

HAMILTON SMITH, Jr., representing North Bloomfield G. M. Co. and Badger Hill M'g Co.

HAMILTON SMITH, Jr.
Chairman Temporary Organization.

NOTES

CHAPTER 1: THE BEGINNING

[1] J. D. Borthwick. *Three Years in California*, 138.
[2] E. Gould Buffum, *Six Months in the Gold Mines*, John W. Caughey, ed., 45.
[3] Rodman W. Paul, *California Gold*, 49.
[4] Ibid, 55.
[5] W.W. Staley, "Elementary Methods of Placer Mining," 7-8.
[6] John Walton Caughey, *Gold is the Cornerstone*, 161.
[7] William Kelly, *A Stroll Through the Diggings of California*, 13-14.
[8] *Gold Mines and Mining in California*, 55-56.
[9] Buffum, 46.
[10] John Bidwell, "The Discovery of Gold in California", 531.
[11] James Mason Hutchings, *The Miner's Own Book*, 3.
[12] E. G. Waite, "Pioneer Mining in California," 130.
[13] Buffum, 38.
[14] Hutchings, 4-5.
[15] Ibid., 6-7; John S. Hittell, *The Resources of California*, 243-245.
[16] Ibid., 245.
[17] Ibid., 250-251
[18] Ibid., 250.
[19] J. Ross Browne, *Reports Upon the Mineral Resources of the United States*, 22.
[20] Walter Van Tilburg Clark, ed., *Journal of Alfred Doten, 1849–1903*, Vol.1, 178.
[21] Ibid., 182.
[22] Paul, *California Gold*, 60.
[23] Charles G. Yale, "The Mineral Industry of California," 34.
[24] Titus Fey Cronise, *The National Wealth of California*, 537. *Gold Mining in California*, 60-61.
[25] Buffum, 61.
[26] Charles D. Ferguson, *California Gold Fields*, 88.
[27] John W. Audubon, *Audubon's Western Journal: 1849-1850*, 221.
[28] Yale, "The Mineral Industry of California," 34.
[29] Edwin F. Bean, *Bean's History and Directory of Nevada County, California*, 30-65; Hubert Howe Bancroft, *History of California*, Vol. VI, 258-359.
[30] Lorenzo Sawyer, *Way Sketches*, 118-120.
[31] Helen S. Griffen, ed. *The Diaries of Peter Decker, Overland to California in 1849 and Life in the Mines, 1850-1851*, 233.
[32] Ibid., 295
[33] Borthwick, 155.
[34] Yale, "The Mineral Resources of California," 20.
[35] Edwin G. Gudde and Elizabeth K. Gudde, ed. *California Gold Camps*, 86-87.
[36] Audubon, *Audubon's Western Journal*, 212.
[37] Ferguson, 124.
[38] Thompson and West, *History of Nevada County, California*, 171.
[39] Ibid., 172.
[40] Philip Ross May, *Origins of Hydraulic Mining in California*, 49.
[41] Clark, *Journal of Doten*, Vol.1, 182-183.
[42] C.A. Logan, "History of Mining and Milling Methods in California," 31.
[43] Hutchings, 7-8.
[44] *Gold Mines and Mining*, 113-114.
[45] R. H. Stretch, *Iowa Hill Canal and Gravel Mines*, 5-6.
[46] Paul, *California Gold*, 148-149.
[47] Olaf P. Jenkins, "Geology of Placer Deposits,"

4.
[48]Ibid., 149.
[49]Nevada City *Daily Transcript*, Nov. 28, 1878, p. 2. Hereafter cited only as *Transcript*.
[50]Paul, 48-49.
[51]For a discussion of this issue see the Nevada County Historical Society *Bulletin*, Vol. 49, No.1, (January 1995) 6.
[52]Rossiter W. Raymond, *Mineral Resources in the States and Territories West of the Rocky Mountains, (1873)* 390.
[53]Ibid., 390-391.
[54]Julian Dana, *The Sacramento River of Gold*, 145-146.
[55]May, 73-74.
[56]Logan, 31.
[57]*Mining and Scientific Press* (San Francisco) Dec. 30, 1893. Hereafter cited as M&SP.
[58]Hutchings, 25.
[59]May, 67-69.
[60]Ibid., 46; J. Ross Browne, 22.
[61]Ibid.,45.
[62]*Transcript*, May 23, 1875, 3.
[63]Hutchings, 25.
[64]Olaf P. Jenkins, "Report Accompanying Geologic Map of Northern Sierra Nevada," Chapter of *Report XXVIII* of the State Mineralogist, (Sacramento: 1932) 287.
[65]Eugene B. Wilson, *Hydraulic and Placer Mining*, 12.
[66]Jenkins, "Geology of Placer Deposits," 6
[67]Charles Scott Haley, "Gold Placers of California," 119-120.
[68]Ibid., 12.
[69]Benjamin Silliman, *Report on the Deep Placers*, 5.
[70]Paul, 40.
[71]Taliesin Evans, *Hydraulic Gold Mining in California*, 7.
[72]Waldemar Lindgren, "The Tertiary Gravels of the Sierra Nevada of California," 71.
[73]Ibid., 65.
[74]Ibid., 65.
[75]Ibid., 66.
[76]C.E. Julihn and F.W. Horton, "Mineral Industries Survey of the United States: California. Tuolumne and Mariposa Counties. Mother Lode District (South)," 158.
[77]Ibid., 160.
[78]C.A. Logan and Herbert Franke, "Mines and Mineral Resources of Calaveras County," 325.
[79]Haley, "Gold Placers of California," 19.
[80]Raymond, 1873, 55.
[81]Lindgren, 133; *Transcript*, August 22, 1874, 2.

CHAPTER 2: THE EARLY YEARS

[1]Taliesen Evens, *Hydraulic Gold Mining in California*, 4; Nevada *Daily Transcript*, May 23, 1875, 2; C.A. Logan "History of Mining and Milling Methods in California, 31; Rossiter W. Raymond, *Statistics of Mines and Mining in the Western States and Territories West of the Rocky Mountains*, (Washington, 1873) 390; H.G. Hanks, "Placer, Hydraulic and Drift Mining," 67-68; Phillip Ross May, *Origins of Hydraulic Mining in California*, 65.
[2]Hanks, 68.
[3]Ibid., 54; John S. Hittell, *Mining in the Pacific States of North America*, 145.
[4]Bowie, *Practical Treatise on Hydraulic Mining*, 49.
[5]Ibid., 180.
[6]Hal Goodyear, "Giants of the Gold Rush," 54; *Transcript*, April 26, 1882, 3.
[7]Nevada City *Gazette*, June 30, 1868, 1.
[8]Joshua Hendy Machine Works, *Hendy Hydraulic Giants*, (San Francisco: 1922) 4.
[9]John S. Hittell, *The Resources of California*, 254.
[10]Robert L. Kelley, *Gold vs. Grain*, 28.
[11]Rodman Paul, *California Gold*, 155.
[12]M.&S.P., December 30, 1893, 422-423.
[13]Sacramento *Daily Union*, July 15, 1854, 1.
[14]J. Ross Browne and James W. Taylor, *Reports on the Mineral Deposits*, 23.
[15]Edwin F. Bean, *Bean's History and Directory of Nevada County*, 131.
[16]Letter Sheet, History Room, California State Library, Sacramento.
[17]Placer *Herald*, January 30, 1855, p. 3.
[18]Sacramento *Daily Union*, July 15, 1854, p. 4.
[19]John S. Hittell, *Mining in the Pacific States*, 145.
[20]James J. Sinnott, *History of Sierra County, Volume V "Over North,"* 181-182.
[21]Harry L. Wells and W. L. Chambers, *History of Butte County*, 209; Rosena A. Giles, *Shasta*

NOTES

County California, (Oakland: Biobooks, 1949) 62; Helen Weaver Gould, La Port Scrapbook, (La Port: Privately printed, 1972) 4-5.

[22] Calaveras Weekly Chronicle, September 10, 1853, 3.
[23] Columbia Gazette, April 20, 1856, p. 1.
[24] Browne and Taylor, 24.
[25] Grass Valley Union, December 14, 1857, p. 1.
[26] Thompson & West, History of Amador County, California, (Oakland: 1881) 208-209.
[27] Judson Farley, "Yuba Hydraulic Mines," 216-218.
[28] Hittell, Resources, 144.
[29] Hanks, 56; Silliman, 19; Titus Fey Cronise, The Natural Wealth of California, 545; M&SP, December 28, 1860, 1.
[30] Silliman, 16-17.
[31] Ibid., 30-31.
[32] Ibid., 35.
[33] Thompson and West, Nevada County, 172.
[34] Raymond, Statistics, 1873, 397.
[35] Samuel Bowles, Our New West, Records of Travel between the Mississippi River and the Pacific Ocean, 420.
[36] Raymond, (1869), 32.
[37] Bean, 378.
[38] Ibid., 396.
[39] August 10, 1861, 4.
[40] Paul, 145.
[41] Charles D. Yale, "The Mineral Industry of California," 8-9.
[42] John S. Hittell, "Mining Excitements in California," 414-415.
[43] M & SP, May 11, 1863, 4.
[44] Thompson and West, Nevada County, 59.
[45] Nevada City Gazette, April 23, 1866, 3.
[46] M&SP, September 14, 1863, 4.
[47] John S. Hittell, The Resources of California, 439-440.
[48] Nevada, Transcript, March 14, 1875, 2.
[49] Ralph Mann, After the Gold Rush, 29
[50] Sinnott, History of Sierra County, "Over North," Vol V, 116.
[51] Grass Valley Union, November 22, 1865, 3.
[52] Transcript, March 22, 1866, 3; April 2, 3.
[53] Grass Valley Daily Union, January 10, 1865, 3.
[54] M&SP, February 11, 1865, 86.
[55] Ibid., December 2, 1865, 342.
[56] Ibid., March 17, 1866, 164.
[57] Transcript, April 5, 1866, 3.
[58] Ibid., April 16, 1866, 3.
[59] Ibid., May 4, 1866, 3.
[60] Ibid., April 6, 1866, 3.
[61] Ibid., October 20, 1866, 3.
[62] Ibid., October 18, 1866, 3.
[63] Ibid., April 25, 1866, 3.
[64] Browne and Taylor, 65; Hanks, 70.
[65] Grass Valley Union, January 5, 1865, 3; Raymond (1869) 30.
[66] Browne and Taylor, 23
[67] Thompson & West, 180.
[68] Grass Valley Union, August 5, 1865, 3.
[69] March 13, 1866, 3.
[70] March 12, 1865, 3.
[71] Browne and Taylor, 65
[72] Transcript, March 8, 1866, 3; also March 14, 1866, 3; Raymond, 70; Earl Mac Boyle, Mines and Mineral Resources of Nevada County, 108.
[73] March 3, 1866, 3.
[74] Ibid., 3.
[75] Bean, 60.
[76] Cronise, 540.
[77] Bean, 53.
[78] Transcript, April 6, 1866, 3.
[79] Ibid., March 14, 1866, 3.
[80] Ibid., May 29, 1866, 3.
[81] Ibid., May 27, 1866, 3.
[82] Downieville Mountain Messinger, April 25, 1874, 3.
[83] Transcript, May 2, 1866, p. 3; May 9, 1866, 3.
[84] Joshua Hendy, 14-15.
[85] Farley, 215.
[86] Cronise, 571.
[87] Transcript, April 6, 1866, 3.
[88] Ibid., May 8, 1866, 3.
[89] Cronise 585-586.
[90] Transcript, June 10, 1874, 3.
[91] Daily National Gazette, June 23, 1870, 3; Transcript, January 19, 1872, 3; Gold Mines of California, 106.

CHAPTER 3: WATER, DITCHES AND DAMS

[1] Nevada Daily Transcript, March 31, 1871, 3.
[2] Ibid., March 29, 1872, 3.

³Ibid., August 18, 1876, 2.
⁴James J. Sinnott, History of Sierra County, "Over North", Vol. V, 238.
⁵Transcript, April 25, 1866, 3.
⁶Ibid., May 15, 1866, 3.
⁷Haley, "Gold Placers of California," 41.
⁸Nevada Daily Gazette, July 25, 1868, 1.
⁹Sacramento Daily Union, October 25, 1855, 2.
¹⁰Ibid., May 4, 1854, 3.
¹¹February 8, 1861, 1.
¹²Edna Bryan Buckbee, The Saga of Old Tuolumne, 266-268; Dane, Ghost Town, 58-59.
¹³J. Ross Browne, Mineral Resources, 1867, 17.
¹⁴Bowles, Our New West, 423.
¹⁵Silliman, Report on the Deep Placers, 30.
¹⁶Shinn, Mining Camps, 250-251.
¹⁷H. P. Davis, Gold Rush Days in Nevada City, 28.
¹⁸Transcript, June 24, 1874, 2; June 25, 2.
¹⁹Ibid., July 4, 1876, 2.
²⁰Josiah Royce, California From the Conquest, 316.
²¹Decker, The Diaries of, 226-227.
²²John S. Hittell, Mining in the Pacific States, 176, 187.
²³Gardiner, In Pursuit of the Golden Dream, 140-242.
²⁴Norris Hundley, Jr. The Great Thirst, California and Water 1770s-1990s, 74.
²⁵Ibid. preface.
²⁶Hittell,189.
²⁷Sinnott, History of Sierra County, Downieville, Gold Town on the Yuba, Vol. I, 229.
²⁸Thompson & West, History of Nevada County, 173.
²⁹Charles Volney Averill, "Placer Mining for Gold in California," 96.
³⁰E. D. Gardner and C. H. Johnson, "Placer Mining in the Western United States, 10; Bowie, A Practicle Treatise on Hydraulic Mining in California, 145.
³¹Wilson, Hydraulic and Placer Mining, 68.
³²Averill, "Placer Mining for Gold in California," 97.
³³Letter from Henry Pichoir to V. G, Bell, Milton Mining and Water Company, Manuscript collection, California State Library.
³⁴Transcript, October 31, 1879, 2.
³⁵Haley, "Gold Placers of California," 41.

³⁶Gold Mines and Mining, 94.
³⁷History and Heritage Committee, Sacramento Section, American Society of Civil Engineers, Historic Civil Engineering Landmarks of Sacramento and Northeastern California, (Sacramento: 1976) 7-8.
³⁸Gold Mines and Mining, 93-95; Gardner & Johnson, 12; Taliesin Evans, Hydraulic Gold Mining in California, 10-11; Bowie, 145.
³⁹Transcript, April 4, 1883, 3.
⁴⁰Ibid., March 11, 1874, 2.
⁴¹Ibid., April 8, 1876, 2.
⁴²Ibid., April 9, 1876, 2.
⁴³Ibid., December 28, 1879, 2.
⁴⁴Grass Valley Union, July 19, 1865, 3.
⁴⁵Bowie, 97-99.
⁴⁶Transcript, August 30, 1882, 3.
⁴⁷Gold Mines and Mining, 98-99.
⁴⁸Thompson & West, 173; Transcript, July 16, 1876, 1; Bean 70.
⁴⁹August 13, 1859, 3.
⁵⁰J. W. Johnson, "Early Engineering Centers in California," 194-195.
⁵¹Silliman, 27-28; Hittell, 80; Bean, 65-66; Thompson & West, 183.
⁵²Transcript, July 25, 1883, 3.
⁵³John S. Hittell, Mining in the Pacific States of North America, 90.
⁵⁴Rossiter W. Raymond, Statistics of Mines and Mining, 45.
⁵⁵Ibid., 52.
⁵⁶Waldemar Lindgren, The Tertiary Gravels of the Sierra Nevada, 173.
⁵⁷Raymond, 1873, 46.
⁵⁸M&SP, January 23, 1875, 57.
⁵⁹California Miners Assn., California Mines and Minerals, 29, 307.
⁶⁰Raymond, 57; Jack Wagner, Gold Mines of California, 27.
⁶¹Raymond, 1873, 58.
⁶²M&SP, February 19, 1876.
⁶³R. H. Stretch, Iowa Hill Canal and Gravel Mines, Placer County, California, 11-12.
⁶⁴Ibid., 23-24.
⁶⁵Ibid., 7.
⁶⁶Raymond, 1873, 79.
⁶⁷Sinnott, History of Sierra County, "Over North" in Sierra County, Vol. V, 9.
⁶⁸William B. Clark, Gold Districts of California,

Bulletin 193, California Division of Mines and Geology, (Sacramento: 1976) 86; Raymond, 1873, 87; Fariss and Smith, *History of Plumas, Lassen & Sierra Counties, 1882,* 480-481.

[69]Sinnott Vol. 5, 233.

[70]Ibid., 151, 156, 477, 480.

[71]Clark, 86.

[72]*M&SP*, January 23, 1875, 57; Farriss & Smith, *History of Plumas, Lassen & Sierra Counties, California, 1882,* 319; Raymond, 1873, 86.

[73]Litchfield, "Smartsville, Timbuctoo and Vicinity," *The History of Yuba County,* 59-61; Raymond, 1873, 70.

[74]J. Ross Browne and James W. Taylor, *Reports upon the Mineral Resources of the United States,* 68-69; Raymond, 1873, 70; Cronise, *The Natural Wealth of California,* 585.

[75]Gardner, 45; Walter W. Bradley, *California Journal of Mines and Geology, Report 38,* (San Francisco: January 1942) 24; Averill, 25.

[76]Harry L. Wells and W. L. Chambers, *History of Butte County,* 211-212.

[77]Ibid., 211-212; Raymond, 1873, 72.

[78]Wells, 209.

[79]Thompson & West, *History of Amador County, California, 1881,* reproduction copy, (Oakland: California Traveler, no date listed) 263-64.

[80]Ibid., 266; *M&SP*, January 23, 1875, 57; also, December 18, 1875, 388; Gardner and Johnson, 41.

[81]William Ireland, Jr., Eighth Annual *Report* of the State Minerologist, (Sacramento: State Mining Bureau, 1888) 148-149; William B. Clark and Philip A Lydon, "Calaveras County, California," 77; C.A Logan and Herbert Franke, Mines and Mineral Resources of Calaveras County," 225-237.

[82]Charles G. Yale, "The Mineral Industry of California," 331, 342.

[83]*M&SP*, Jan. 23, 1875, 57.

[84]Jack I. N. Brotherton, *Annals of Stanislaus County,* (Santa Cruz: Western Tanager Press, 1982) 166.

[85]See John W. Robinson, *The San Gabriels,* (Arcadia, Calif., Big Santa Anita Historical Society, 1991); Frank S. Wedertz, *Mono Diggins,* (Bishop, Calif., Chalfant Press, 1978); Raymond, 1873, 99.

[86]See Principal Hydraulic Mines, Appendix B, from William B. Clark, *Bulletin 193.*

[87]*Transcript,* March 11, 1871, 3.

[88]Ibid., March 17, 1871, 3.

[89]Ibid., March 25, 1871, 3.

[90]Ibid., 2.

[91]Ibid., August 23, 1874, 2; Charles G. Yale, "The Mineral Industry of California," 28.

[92]Ibid., 27.

[93]*Transcript,* August 28, 1875, 2.

[94]Ibid., May 12, 1880, 2.

[95]Advertisement in *M&SP,* February 1, 1861, 7.

[96]Bowie, 49.

[97]Ibid. 158-160.

[98]Wilson, *Hydraulic and Placer Mining,* 53-55.

[99]Robert H. Richards, *Mining Engineering Notes,* Vol. I, 251.

[100]Cronise, 609-615.

[101]Bowie, 79, 166; Wilson, 63; Alice Goen Jones, editor, *Trinity County Historical Sites,* 187.

[102]Evans, 12; Jack R. Wagner, *Gold Mines of California,* (Berkeley: Howell-North, 1970) 27.

CHAPTER 4: INVENTION AND MATURITY

[1]Rossiter W. Raymond, *Statistics of Mines and Mining,* 81.

[2]Ibid., 7.

[3]Samuel Bowles, *Our New West,* 420.

[4]Charles G. Yale, "The Mineral Industry of California," 2.

[5]These figures were compiled by Rodman W. Paul from a variety of sources see his *California Gold,* 349-350.

[6]Nevada *Daily Transcript,* May 18, 1879, 2.

[7]Yale, 263.

[8]Quoted from the *Alta* by the Nevada *Daily Transcript,* July 24, 1867, 3.

[9]Thomas Starr King, quoted from the Boston *Transcript* by the Nevada City Democrat, January 3, 1861; Nevada *Daily Transcript,* June 23, 1874, 3. Thomas Starr King was a reknowned Unitarian Minister, lecturer and orator. He appeared at the Metropolitan Thearer in Nevada City in 1860. He wrote letters about his impressions of the gold region which were printed by the Boston

paper.

[10] Nevada Daily National Gazette, April 29, 1870, 3.

[11] Transcript, November 29, 1871, 3.

[12] R. H. Stretch, Iowa Hill Canal and Gravel Mines, Placer County, 6-7.

[13] Gold Mines and Mining in California, 115.

[14] Joshua Hendy Iron Works, Hendy Hydraulic Giants, 30-33.

[15] Transcript, April 26, 1882, 3.

[16] Nevada Daily Gazette, September 18, 1869, 3.

[17] Aug. J. Bowie, A Practical Treatise on Hydraulic Mining in California, 180.

[18] Gazette, September 18, 1869, 3.

[19] Ibid., November 14, 1869, 3.

[20] Ibid., April 14, 1870, 3.

[21] Ibid., May 17, 1870, 3.

[22] M&SP, October 1, 1870, 240.

[23] Gazette, July 20, 1870, 3.

[24] Ibid., October 5, 1878, 2.

[25] Bowie, 181.

[26] Gazette, October 1, 1870, 3.

[27] Bowie, 182-183.

[28] Quoted by the Transcript, July 1, 1882, 3.

[29] Transcript, April 26, 1882, 3.

[30] H. G. Hanks, "Placer, Hydraulic and Drift Mining," 70.

[31] Transcript, April 18, 1876, 2.

[32] Hanks, 66.

[33] Manuscript at the California State Library, Box 857 of the Milton Mining & Water Company.

[34] Ibid., Letter from Henry Pichoir to V. G. Bell.

[35] Bowie, 50.

[36] Ibid., 183.

[37] Hendy, 4.

[38] E. D. Gardner and C. H. Johnson, "Placer Mining in the Western United States, Part II, Hydraulicking, Treatment of Placer Concentrates, and Marketing of Gold." Information Circular, United States Bureau of Mines, (Washington: October 1934) 21.

[39] Hendy, 5.

[40] Ibid., 58-59.

[41] E.D. Gardner, 38.

[42] Nevada Daily National Gazette, July 18, 1870, 3.

[43] Eugene B. Wilson, Hydraulic and Placer Mining, 72; Hendy, 17.

[44] Dennis H. Stovall, "Art of Placer Mining," Mining and Scientific Press, November 13, 1909, 661-662.

[45] Wagner, Gold Mines, 27.

[46] May 13, 1876.

[47] Bowie, 245, Robert H. Richards, Mining Engineering Notes, Vol. I., 251.

[48] Hendy, 17.

[49] James Mason Hutchings, The Miner's Own Book, 27.

[50] Grass Valley Union, January 15, 1865.

[51] W. W. Kallenberger, Memories of a Gold Digger, 10-11.

[52] Transcript, September 13, 1878, 2.

[53] Wilson, 70-71.

[54] Charles Scott Haley, "Gold Placers of California," 37-38.

[55] Hendy, 36.

[56] Hal Goodyear, "Giants of the Gold Rush," 54, Taliesin Evans, Hydraulic Gold Mining in California, 9-10.

[57] Stovall, 661-662.

[58] E.D. Gardner, 39. Charles Volney Averill, "Placer Mining For Gold in California," 112-114. Hanks, 111.

[59] Richards, 259; Charles G. Yale, "The Mineral Industry of California," 31.

[60] Hendy, 11.

[61] Hutchings, 27; Gardner, 3-4.

[62] Stovall, 662.

[63] Haley, "Gold Placers of California," 40.

[64] Wilson, 93.

[65] Transcript, June 10, 1876, 2.

[66] Bowie, 225-227.

[67] M&SP, July 6, 1861, 6.

[68] Ibid., March 28, 1914.

[69] Haley, 42-43.

[70] Cronise, 546; Bowie, 227-228.

[71] Hanks, 112; Bowie, 144-145.

[72] Richards, 259-260.

[73] Bowie, 248; Averill, 130.

[74] Averill, 139.

[75] Hanks, 113.

[76] Evans, 8-9.

[77] Transcript, September 8, 1878, 2.

[78] Grass Valley Union, August 1, 1865, 3.

[79] Transcript, March 11, 1866, 3.

[80] Ibid., February 15, 1865, 3.

[81] Ibid., August 30, 1866, 3.

[82] Cronise, 572; W.W. Staley, "Elementary Methods of Placer Mining," 12.
[83] M&SP, March 13, 1875, 168; Ibid., May 13, 1876; Bowie, 231; Gardner, 74; Hanks, 87.
[84] M&SP, August 17, 1861, 1.
[85] Transcript, October 27, 1866, 3.
[86] Thompson & West, 180.
[87] Sinnott, History of Sierra County, "Over North" Vol V, 235.
[88] Nevada Gazette, June 19, 1869, 3.
[89] Richards, 259.
[90] Bowie, 244.
[91] M&SP, March 9, 1895, 154.
[92] Richards, 259.
[93] Thompson & West, 180.
[94] F.W. Robinson, "Notes on Hydraulic Mining," 127.
[95] Transcript, April 4, 1880, 2.
[96] Hanks, 39.
[97] John S. Hittell, The Resources of California, 248.
[98] John Hays Hammond, "Auriferous Gravels in California," 132.
[99] Bowie, 176.
[100] Hanks, 50; Bowie, 178.
[101] Bowie, 178-179.
[102] DeGroot, "Hydraulic and Drift Mining," 66.
[103] Transcript, March 14, 1876, 3.
[104] Averill, 113-114.
[105] Bowie, 185.
[106] M&SP, August 3, 1861, 4.
[107] Robert J. and Grace J. Slyter, Historical Notes of the Early Washington, Nevada County, California Mining District, (Privately Printed, no date) 108.
[108] M&SP, August 3, 1861, 4.
[109] Ibid., July 15, 1876, 312.
[110] Joshua Hendy Machine Works, "Gold and Silver Quartz Mining and Milling Machinery," Cataloque No. 2. (San Francisco: 1998) 144.
[111] Grass Valley Union, January 5, 1865, 3.
[112] Transcript, May 2, 1866, 3.
[113] Powell Greenland, "Knight's Foundry," 4.
[114] W. Turrentine Jackson, "Report on the Malakoff Mine, the North Bloomfield Mining District and the Town of North Bloomfield," 51; Bowie 247.
[115] Ibid., 240; Hanks, 111; Transcript, April 9, 1879, 2.
[116] Transcript, May 23, 1875, p. 2.
[117] Ibid., November 29, 1876, 3.
[118] Thompson and West, 61; Transcript, September 25, 1878, 2; also November 7, 1878, p. 3; Wyckoff, "Hydraulicking," 10-11.
[119] California Mines and Minerals, 434-435.
[120] Lardner and Brock, 183.
[121] Richards, 255; Gardner, 23; Wilson, 81-82; Haley, 43-44.
[122] Ibid., 45; E.D. Gardner, 23.
[123] Ibid., 57-58.
[124] M&SP, November, 13, 1909, 662.
[125] Haley, "Gold Placers of California," 40.
[126] Gold Mines of California, 109-111; Transcript, April 9, 1876, 2.
[127] Sinnott, Vol. V, 156-157.
[128] Raymond, 1873, 67.
[129] M&SP, September 4, 1869, 150.
[130] Transcript, May 23, 1875, 2.
[131] California Mines and Minerals, 296.
[132] Wilson, 17-18; Bowie, 88.
[133] Lindgren, 71.
[134] Gold Mining in California, 115.
[135] Haley, 9; Democrat, January 3, 1861, p. 3; John Hays Hammond, "Auriferous Gravels of California," 132.

Chapter 5: The Summit

[1] Located at the present town of Anderson, Reading was granted 6 leagues in 1844. This was the most northerly of all Mexican land grants. See Robert G. Cowan, Ranchos of California, 73.
[2] Tenth Annual Report of the State Mineralogist, (1890) 695.
[3] Mildred Brooke Hoover, Hero Eugene Rensch, Ethel Grace Rensch, Historic Spots in California, (Stanford: Stanford University Press, 1966) 554.
[4] Bancroft, History of California, Vol. VI, 371.
[5] Susan Sheppard, "La Grange Mine Trinity County," 1.
[6] Isaac Cox, Annals of Trinity County, 9.
[7] Trinity Journal, January 21, 1860, 3.
[8] Sheppard, 1.
[9] Rossiter W. Raymond, Statistics of Mines and Mining, 99.
[10] Hal Goodyear, "Peter Minert Paulsen," 34.

[11] M&SP, May 2, 1874, 278.
[12] Ibid., April 6, 1872, 212.
[13] Quoted from the Trinity Journal by Ibid., May 2, 1874, 278.
[14] Ibid., April 24, 1875, 268; June 5, 1875, 180.
[15] Ibid., Dec. 11, 1875, 373.
[16] Ibid., June 23, 1877, 397.
[17] Ibid., May 18, 1878, 309.
[18] Ibid., June 22, 1878, 388; August 8. 1878, 84.
[19] Goodyear, "Peter Paulsen," 35.
[20] "Trinity Gold Mining Company, Eighth Annual Report of the State Minerologist, (Sacramento: 1888) 638.
[21] M&SP, August 11, 1888, 9.
[22] Tenth Annual Report of the State Mineralogist (Sacramento: 1890) 702
[23] Ibid. 702.
[24] Ibid. 702.
[25] Jake Jackson, Tales From the "Mountaineer." 59.
[26] M&SP, January 11, 1896, 30-31.
[27] Twelfth Annual Report, 311
[28] M&SP, 60; M&SP, August 26, 1893, 142.
[29] Twelfth Annual Report, 311
[30] M&SP, April 28, 1894, 269.
[31] Ibid., August 10, 1895, 94; December 28, 1895, 430.
[31] Jake Jackson, 61.
[33] J. J. Crawford, "La Grange," State Mineralogist Report XIII, (Sacramento: 1896) 452-453.
[34] Vernon Ryan, "La Grange Mine," Yearbook, Trinity County Historical Society, (Weaverville: 1969) 26-29; M&SP, October 10, 1908; 594. Ibid., August 18, 1900, 191.
[35] M&SP, December 11, 1897, 555.
[36] Jake Jackson, 111; Sheppard, 11.
[37] Vernon Ryan, 32.
[38] Joshua Hendy Iron Works, Hendy Hydraulic Giants, (San Francisco: 1922) 30; United States Geological Survey Bulletin No. 430, (1909), 53.
[39] E.D. Gardner and C.H. Johnson, "Placer Mining in the Western United States," 39.
[40] M&SP, January 24, 1903, 60.
[41] Ibid., March 2, 1912, 354.
[42] Ibid., August 18, 1900, 191.
[43] Ryan, 34; Sheppard, 11.
[44] Jake Jackson, 66.
[45] Ibid. 66; J. C. O'Brien, "Mines and Mineral Resources of Trinity County California," 12-13.
[46] M&SP, August 18, 1900, 191; Ibid., July 30, 1910, 101.
[47] State Mining Bureau, Report XIV, 908.
[48] Ibid., April 22, 1905, 257.
[49] Sheppard, 8-9.
[50] M&SP, June 17, 1911, 831.
[51] U. S. Bulletin 430, 53.
[52] M&SP, October 25, 1913, 664.
[53] Ibid., January 10, 1914, 119; February 14, 1914, 308.
[54] Ibid., July 30, 1910, 101.
[55] Ibid., same issue; October 10, 1908, 492.
[56] U. S. Bulletin 430, 53-56.
[57] Untitled manuscript at the Weaverville Museum Research Center; M&SP, January 27, 1917, 141.
[58] Charles Scott Haley, "Gold Placer of California," 94.
[59] Ms. At Weaverville Museum Research Center.
[60] Charles Volney Averill, "Placer Mining for Gold In California," 308.
[61] Ryan, 32.
[62] Raymond, 72.
[63] Jack D. Sturgeon, "Cherokee Flat," Butte County Historical Society Quarterly Diggins, Vol. X:4, (Winter 1966) 11,12.
[64] Harry L. Wells and Chambers, History of Butte County, 1881, reproduction copy (Berkeley: Howell-North, 1973) 212-213.
[65] Sturgeon, 13.
[66] M&SP, March 18, 1871, 164.
[67] Ibid., January 7, 1871, 1-2; Jack D. Sturgeon, 17; Charles Waldeyer quoted in Raymond, Statistics, 1873, 410.
[68] Hugh A. Shamberger, "The Story of the Water Supply For the Comstock . . .," 21.
[69] Quoted in Raymond, Statistics, 1873, 391.
[70] Butte Record, December 24, 1870, 3.
[71] M&SP, August 23, 1873, 122.
[72] Copied from the Record, by the M&SP, November 8, 1873, 290.
[73] Sturgeon, 28.
[74] William Ireland, Jr., "Spring Valley Hydraulic Mine," State Mineralogist Sixth Annual Report, (Sacramento: 1886) 24-25.
[75] M&SP, August 29, 1874, 137; Sturgeon, 23.
[76] M&SP, January 10, 1874, 20.
[77] Wells and Chambers, 212.

[78] Rossiter W. Raymond, *Statistics*, Eighth Annual Report (Washington: 1877) 97-100.

[79] Ibid., 213; *M&SP*, May 15, 1875, 318; December 4, 1875, 356; September 8, 1906, 297.

[80] Quoted in the Nevada *Daily Transcript*, July 24, 1880, 2.

[81] Sturgeon, 19.

[82] Wells and Chambers, 213, *M&SP*, September 8, 1906, 297.

[83] Sturgeon, 20.

[84] Waldemar Lindgren, *The Tertiary Gravels of the Sierra Nevada*, 87.

[85] *Transcript*, June 7, 1883, 3.

[86] *M&SP*, December 3, 1887, p. 360; December 10, 1887, 376.

[87] Ibid., December 15, 1888, 400.

[88] Sixth *Report*, 25.

[89] *M&SP*, October 26, 1889, 320.

[90] Ibid., September 8, 1906, 296-297.

[91] J. A. Miner, "Spring Valley Hydraulic Gold Mine," 10th *Report*, (Sacramento: 1891) 124-125.

[92] Lindgren, 87; William B. Clark, "Gold Districts of California," Bulletin 193, California Division of Mines and Geology, (Sacramento: 1976) 19; Robert L. Kelley, *Gold Vs Grain, The Hydraulic Mining Controversy in California's Sacramento Valley*, (Glendale: Arthur H. Clark Co., 1959) 64.

[93] Edwin F. Bean, *Bean's History and Directory of Nevada County*, 395.

[94] Edwin G. Gudde, *California Gold Camps*, 162-163.

[95] Thompson and West, *History of Nevada County, California*, 59, 183; Michel Janicot, *The French Connection or the French in Nevada County*, 13, 15.

[96] Ibid., 15.

[97] W. Turrentine Jackson, *Report on the Malakoff Mine, the North Bloomfield Mining District, and the Town of North Bloomfield*, (Sacramento: Department of Parks and Recreation, 1967).

[98] Janicot, 17-18; Jimmie Schneider, *Quicksilver, The Complete History of Santa Clara County's New Almaden Mine.* (San Jose: Privately printed, 1992) 39. Pioche, a close friend of Butterworths', for a time, leased the lavish home at New Almaden called Casa Grande. Tragically, both Pioche and Poquillon ended their lives by suicide.

[99] Bean, 395; *Transcript*, June 28, 1866, 3; Ibid., July 13, 1866, 3; Ibid., August 10, 1866, 3.

[100] *Transcript*, Sept. 28, 1866, 3.

[101] Grass Valley *Union*, May 15, 1867, 3.

[102] Nevada *Daily Gazette*, July 8, 1868, 1; Ibid., August 14, 1868, 1; *M&SP*, March 7, 1868, 150.; Ibid., November 7, 1866, 294.

[103] *Transcript*, August 25, 1869, 2

[104] Ibid., December 16, 1869, 2.

[105] *Gazette*, August 26, 1869, 3.

[106] *Transcript*, December 16, 1869, 2.

[107] Ibid., July 19, 1879, 2.

[108] *M&SP*, May 22, 1869, 326; also May 29, 1869, 342; *Gazette*, October 10, 1869, 3; Ibid., November 23, 1869, 3.

[109] Ibid., August 1, 1870, 3.

[110] T. Jackson, 15-17.

[111] *M&SP*, October 22, 1870, 282.

[112] Ibid.. October 15, 1870, 268.

[113] *Transcript*, April 4, 1871, 3.

[114] Bowie, *A Practical Treatise on Hydraulic Mining in California*, 88.

[115] *Transcript*, June 30, 1871, 3.

[116] DeGroot, "Hydraulic and Drift Mining," 163; *Transcript*, March 30, 1871, 3.

[117] Ibid., April 11, 1871, 2.

[118] T. Jackson, 13; *Transcript*, January 6, 1872, 3; Ibid., June 18, 1872, 3; also June 26, 1872, 3; also August 15, 1872, 3.

[119] *Transcript*, April 19, 1872, 3.

[120] Grass Valley *National*, April 28, 1869, 2.

[121] *Transcript*, April 23, 1872, 3.

[122] Ibid., April 24, 1873, 3.

[123] DeGroot, 163-164; *M&SP*, June 30, 1877, 420.

[124] *Transcript*, August 16, 1872, 3; T. Jackson, 29.

[125] Kallenberger, *Memories of a Gold Digger*, 8-9.

[126] *Transcript*, July 19, 1881, 3.

[127] *M&SP*, January 16, 1875, 37.

[128] Charles H. Shinn, quoted from the San Francisco *Bulletin* by the *Transcript*, July 19, 1879, 2.

[129] Hamilton Smith, Jr., "The Bowman Reservoir and Dam," 37-38; *M&SP*, February 15, 1879, 105.

[130] *Transcript*, June 10, 1982, 3.

[131] Smith, "The Bowman Reservoir and Dam,"

105.

[132] *Transcript*, April 18, 1876, 2.
[133] Ibid., May 17, 1876, 2.
[134] *M&SP*, September 30, 1876, 221; November 3, 1877, 277; January 19, 1878, 41; T. Jackson, 49; *Transcript*, June 28, 1881, 2.
[135] *Transcript*, September 5, 1876, p. 2; David Lavender, *Nothing Seemed Impossible*, 2.
[136] Manuscript, Milton papers, California State Library.
[137] T. Jackson, 65.
[138] Ibid., 65.
[139] *M&SP*, November 2, 1878, 277; February 15, 1879, 105.
[140] Hammond, "The Auriferous Gravels of California," State Mineralogists's *Report IX*, (1889) 111-113.
[141] *M&SP*, February 15, 1879, 105.
[142] Ibid., July 3, 1880, 9.
[143] *Transcript*, May 29, 1880, 2.
[144] *Transcript*, July 30, 1879, 2.
[145] Ibid., February 11, 1880, 2.
[146] Ibid., March 10, 1880, 2; *M&SP*, October 14, 1882, 327.
[147] *Transcript*, November 7, 1878, 2.
[148] Ibid., May 31, 1881, 2.
[149] Ibid., June 28, 1881, 2.
[150] Ibid., Oct. 29, 1882, 3; Oct. 31, 3.
[151] *M&SP*, July 2, 1881, 5.
[152] Ibid., October 22, 1881, 269.
[153] *Transcript*, January 22, 1884, 3. For the definitive study on the debris controversy see Robert L. Kelley, *Gold vs. Grain*.
[154] T. Jackson, 91; *Transcript*, August 19, 1883, 3. Johnson, J.W. "Early Engineering Centers in California" Califoarnia Historical Society *Quarterly*, Vol. XXIX, 1950, 200
[155] *Transcript*, February 12, 1894, 2.
[156] Kirk Monroe, "Hydraulic Mining in California," *Harpers Weekly*, No. 39, (1895).
[157] Grass Valley *Union*, Oct. 8, 1893, 3.
[158] Errol Mac Boyle, "Mines and Mineral Resources of Nevada County," 100.

CHAPTER 6: THE ENDING

[1] W. T. Ellis, *Memories, My Seventy-two Years in the Romantic County of Yuba, California*, 145; Charles Scott Haley, "Gold Placers of California," 9.
[2] Ellis, *Memories*, 146.
[3] A. Judson Farley, "Roses Bar," *Overland Monthly*, Vol. VII, (1871).
[4] Nevada City *Daily National Gazette*, July 8, 1870, 3.
[5] *Transcript*, July 18, 1875, 2.
[6] Ibid., March 22, 1876, 2.
[7] Ibid., August 5, 1876, 2.
[8] Manuscript in Milton Collection, California State Library.
[9] *Transcript*, April 20, 1876, 2.
[10] Quoted in the *Transcript*, March 19, 1876, 2.
[11] Ibid., October 3, 1879, 2.
[12] Ibid., March 12, 1879, 2.
[13] Kelley, *Gold vs. Grain*, 115.
[14] Quoted in the *Transcript*, June 29, 1879, 2.
[15] *Transcript*, November 18, 1879, 2.
[16] Ibid., July 9, 1880, p. 2; August 12, 1880, 2.
[17] Quoted in the *Transcript*, July 2, 1881, 2.
[18] Ibid., June 29, 1881, 2.
[19] *Transcript*, August 23, 1879, 2.
[20] Ibid., October 18, 1881, 2.
[21] Quoted in the *Transcript*, June 29, 1881, 2.
[22] Ibid., July 14, 1881, 2.
[23] Ibid., June 21, 1881, 2.
[24] Ibid., June 16, 1881, 1.
[25] Ibid., June 9, 1881, 2.
[26] Kelley, *Gold vs. Grain*, 196-197.
[27] *Transcript*, June 5, 1881, 2.
[28] Ibid., June 8, 1881, 2; June 11, 2.
[29] Quoted in the *Transcript*, July 14, 1881, 2.
[30] Ibid., July 21, 1881, 2.
[31] Ibid., November 20, 1881, 2; Waldemar Lindgren, *The Tertiary Gravel Gravels of the Sierra Nevada of California*, 66; Evans, *Hydraulic Gold Mining in California*, 12.
[32] *Transcript*, November 12, 1881, 2.
[33] Ibid., April 6, 1882, 2.
[34] *California Mines and Minerals*, 413.
[35] *Transcript*, May 3, 1882, 3.
[36] Ibid., May 9, 1882, 3.
[37] *M&SP*, July 1, 1882, 8.
[38] *Transcript*, July 22, 1882, 3.
[39] Ibid., September 22, 1882, 3; September 24, 3; September 26, 3.
[40] *M&SP*, December 16, 1882, 392.
[41] *Transcript*, March 27, 1883, 3.

42 Quoted in the *Transcript*, March 30, 1883, 2.

43 *Transcript*, June 19, 1883, 3; June 20, 3; June 21, 3; June 22, 3; June 23, 3; Ellis, *Memories*, 80; Wycoff, *Hydraulicking a Brief History of Hydraulic Mining in Nevada County, California*, 23; August J. Bowie, Jr., *A Practical Treatise on Hydraulic Mining in California*, 93; Van Der Pas, "The English Dam Disaster, June 18, 1883," 18-19; Doris Foley, "English Dam Catastrophe Hastened Sawyer Decision," Nevada City *Nugget*, Centennial Publication, May 18, 1951, 50.

44 Milton Manuscript file, California State Library, California Room.

45 Marysville *Daily Appeal*, January 8, 1884, 3.

46 *Transcript*, January 11, 1884, 2.

47 *Appeal*, January 8, 1884, 2.

48 Report of the California Debris Commission, November 15, 1893, 4.

49 Grove Karl Gilbert, "Hydraulic Mining Debris in the Sierra Nevada," 38.

50 *Transcript*, February 3, 1887, 3.

51 Ibid., November 20, 1883, 3.

52 Sinnott, *History of Sierra County, Volume V, "Over North,"* 240.

53 *Transcript*, February 17, 1884, 3; September 14, 3; February 13, 3.

54 Sinnott, Vol. V, 240.

55 Quoted by the *Transcript*, December 14, 1884, 3.

56 August J. Bowie, "Mining Debris in California Rivers" Technical Society of the Pacific Coast, *Transactions*, Volume IV, (February 1887) 13.

57 Quoted by the *Transcript*, March 18, 1876, 2.

58 *California Mines and Minerals*, 415.

59 *Gold Mining in California*, 260-262.

60 Ibid., 213, 266.

61 James M. Hill, "Historical Summary of Gold, Silver, Copper, Lead and Zinc Produced in California 1848-1926, 4, 6; Ralph C. Loyd "Gold Mining Activity in California," 169; Arthur Jarman "Report of the Hydraulic Mining Commission, 12.

62 Charles Scott Haley, "Gold Placers of California," 13, 14; Charles G. Yale, "Mining Debris Legislation," *California Mines and Minerals*, (San Francisco: Miner's Association, 1899) 257-259.

63 *Transcript*, April 19, 1895, 3; November 23, 1895, 3; also articles in the following editions: January 24, 26, 31, 1896.

64 Ibid., October 6, 1904, 3; January 2, 1905, 3.

65 Lindgren, *Tertiary Gravels*, 82; Lloyd L. Root, California Hydraulic Mining Commission, *Report* on the Resumption of Hydraulic Mining in California, (1927) 30-33.

66 Quoted in *California Mines and Minerals*, 413.

67 Arthur Jarman, "Report of the Hydraulic Mining Commission Upon the Feasibility of the Resumption of Hydraulic Mining in California," 40-83; Errol Mac Boyle, "Mines and Mineral Resources of Nevada County," 101.

68 Warren E. Yeend, "Gold Bearing Gravel of the Ancestral Yuba River, Sierra Nevada, California," 3.

69 Sacramento *Union*, January 6, 1937.

70 *Hydraulic Miner*, (Grass Valley) Vol. 1, No. 1, March 1939. Suspended operations March, 1941.

71 Yeend, 4.

Glossary of Placer Mining Terms

Adit. A passage or tunnel driven horizontally into a mine from the side of a hill. In cement mining the tunnel was used in conjunction with a cross-cut for blasting purposes.

Amalgam. A combination of quicksilver and gold left in the riffles of a cradle, long tom or sluice after the washing process.

Amalgam kettle. An ordinary sheet-iron bucket or porcelain-lined iron kettle. In the clean-up operation it was used as a receptacle for floating the gold amalgam.

Anchor ice. The build-up of ice on the sides and bottom of a flume during freezing weather.

Auriferous gravel. Containing gold, gold-bearing.

Bank water (by-water or by-wash). A stream of water directed over the bank of a hydraulic mine and into the pit for the purpose of helping wash caved material into the sluices.

Batea. A broad cone-shaped wooden dish used by Mexican and South American miners to pan gold.

Bedrock cut (sometimes referred to as a ground sluice). A trench carved into the bedrock in a hydraulic pit leading from the head of the sluices to the working face or gravel bank.

Bedrock tunnel. A tunnel driven at bedrock level or below, with sufficient grade for use as an outlet for material washed from a hydraulic mine.

Blasting powder. Common gunpowder or black powder used until the late 1860s when dynamite was introduced into the California mines. Dynamite was usually used in combination with black powder for bank or cement blasting.

Blue lead. A misnomer applied by miners who believed there was a single

Tertiary channel running from Mariposa to Plumas County. The rich bottom strata was characterized by a deep bluish color.

Blower. A shallow dish with an opening on one side used to blow away sand to separate the gold.

Booming (self-starter). A method of automatically releasing an abundance of water, in a wave action to facilitate a ground sluicing operation.

Booster (drive giant). A large monitor, usually constructed with no horizontal movement, used to help move heavy material through the sluices.

Caving. The action of a stream of water, from a nozzle, causing a bank to be brought down.

Cement. Compacted soil formed in Tertiary deposits by the superimposed weight of the debris above it and the chemical action of sulfides, making it rock-hard in its consistency.

Charging the sluices. The initial act of dispensing measured amounts of quicksilver into the sluices.

Clean-up. The periodic process of removing the amalgam or black sand and gold from the riffles after the washing operation.

Cox pan. A mechanical device used in cement mills to separate boulders from the finer material.

Coyoting or coyote mining. An early method of mining "dead rivers" (1850) by digging a shaft on the side of a hill to bedrock in much the same manner as a well, then drifting in all directions..

Crevicing (also known as knife mining). An early method of mining the cracks or crevices in rocks, near stream beds, where particles of gold, over the centuries, had accumulated. This was usually accomplished with nothing more than a knife. Also crevicing referred to the final task in a clean-up operation where cracks and nail holes in flumes and riffles were probed for any remaining gold.

Crib dam. A form of wooden dam constructed with notched timbers bolted together in a series of cribs filled with stones. The bottom timbers were bolted to bedrock. The face of the dam was made watertight by an outer skin of planks and batons.

Crinoline hose. Hydraulic hose reinforced with galvanized iron bands.

Crosscut. A horizontal passage or opening leading at right angles from an

GLOSSARY

adit or tunnel used in cement blasting.

Cutting monitor. The nozzle used in bringing down banks as distinguished from the sweeping monitor which swept the material into the sluices.

Dead rivers. A name miners applied to ancient buried stream beds containing auriferous gravel deposits such as those mined by the coyote process.

Debris dam. An obstruction made of brush, wood or dirt placed across beds of streams for the purpose of holding back the sand and gravel (slickens) coming from an hydraulic mine to prevent their entering into the navigable streams and damaging the land in the valleys below.

Deep diggings. Mines of the Tertiary gravel deposits (ancient riverbeds), worked either by drift mining or hydraulic mining.

Deflector. A device designed to guide and facilitate the movement of a monitor in a horizontal direction.

Derrick (bedrock derrick). Composed of a mast held in position by six galvanized iron wire ropes equipped with a boom and whip block. Used for moving large boulders, up to eleven tons, from a mining pit.

Discharge box. A device, used by early water ditch companies, to measure the amount of water sold to customers. It consisted of a long horizontal slot in a board, through which the water flowed. The opening was fitted with a slide to measure and dispense the water.

Distributing box. A cast iron box located at the bottom of a hydraulic claim and supplied with water from a supply line leading from an elevated location. The box was equipped with a gate valve and from four to six couplings to facilitate the connection of crinoline hose.

Distributing pipe. A pipe or hose located at the bottom of a hydraulic claim and connected to a nozzle.

Drawing. Aiming the stream of the nozzle ahead of the debris on the mine floor to cause it to be drawn back toward the nozzle.

Drift. A horizontal underground passage generally following a gold-bearing gravel deposit.

Drift mining (tunnel mining). A method of deep placer mining in Tertiary gravel deposits. It required driving a tunnel at or slightly below bedrock to enter the ancient channel to remove the gold-bearing gravel by wheelbarrow or ore cars.

Drop. A rectangular opening in the farther end of a sluice box, equipped with a grating or grizzly. It was designed for the purpose of letting the finer material drop into an undercurrent placed below the main sluice.

Duty. The amount of gravel, measured in cubic yards, that could be washed by a determined amount of water measured in miner's inches.

Dynamite (giant powder). A powerful explosive composed of nitroglycerin dispersed in an absorbent medium with a combustible dope such as wood pulp.

Face. The surface of the working end of an advancing tunnel.

Flouring. A condition causing quicksilver to break up into minute, dull colored drops that were subsequently lost. This problem was generally the result of improper handling and exposure to air.

Flume. A wooden conduit ranging from two to six feet in width and two to four feet in height to conduct water through shale or shattered ground or along steep cliffs.

Flume tenders or ditch tenders. Workmen employed to observe, protect and maintain the flumes and ditches from robbers, accident and wear.

Gambusino. Mexican or South American miner.

Globe Monitor. An improved hydraulic nozzle with a ball-and-socket joint. It was later further refined by the installation of an interior tripod device.

Gooseneck nozzle. The first major improvement in the hydraulic nozzle. It was designed with two elbows, one working above the other, with a coupling joint in between.

Goosing. The process of sweeping or driving the debris, on the floor of a hydraulic mine, before the stream of a monitor.

Grizzly. A form of grating made of iron bars, set in an opening, designed to prevent heavy material such as boulders, stumps, etc. from passing through.

Ground sluicing. A method of stripping away the overburden, covering a

GLOSSARY

gold-bearing gravel deposit, with a stream of water, .

Hard rock mining (vein or quartz mining). A method of mining where the rock is drilled, blasted and crushed to extract the gold locked in its veins.

Head. A method of expressing the pressure exerted by water in terms of the number of feet of elevation. One hundred feet equals 42.5 psi.

Head of water. The amount of water sold by a ditch company to a customer, for a period of 24 hours or less, measured in miner's inches.

Hendy Giant. An improved monitor, manufactured by the Joshua Hendy Machine Works, using the patent of Frank Fisher's Hydraulic Chief and incorporating a new "king-bolt" design.

Hungarian riffles. A reinforced riffle generally consisting of blocks strengthened with iron or steel strapping.

Hurdy-gurdy. A water wheel used in hydraulic mines to power derricks and electric generators. The name was used to distinguish them from the old fashioned over-shot and undershot wheels in general use. They were technically, tangential impact wheels.

Hydraulic Chief. An improved monitor invented by Frank Fisher. It was also known as the "Knuckle-Joint and Nozzle."

Hydraulic elevator. A device designed to elevate debris from a mine, with no outlet, to a higher elevation.

Hydraulicking. A variant term for hydraulic mining.

Hydraulic mining. The use of a hose and nozzle, under pressure, to wash down a bank and run the material through a sluice.

Inverted siphon. A method of transporting water through an inclined pipe over a steep canyon. The water when entering the pipe must have a higher head than when leaving it.

Lead. A misnomer for lode, vein or a continuous gravel deposit.

Little Giant. A double-jointed hydraulic monitor, invented by Richard Hoskins, which won wide acceptance. It could freely move both horizontally and vertically.

Long tom. A trough usually about twelve feet long equipped with riffles used to wash and save gold from sand and gravel.

Mercury (quicksilver). A heavy silver-white metallic element, liquid at room temperature, used to gather and hold gold, forming an amalgam.

Miner's inch. A method of measuring and charging for water. Although varying somewhat throughout the gold regions of the state, the most widely adopted was the amount of water that could flow through a one-inch orifice in a two-inch plank, six inches below the surface, during a 24-hour period. This quantity was calculated to be 17,000 gallons or 2,274 cubic feet.

Monitor. A high pressure hydraulic nozzle, usually bolted to cross timbers, (not hand held).

Mother lode. A geological and geographic term referring to a continuos gold-bearing quartz vein running for approximately 120 miles north from Mariposa to the vicinity of Georgetown in El Dorado County.

Northern mines. A term used to describe the mines north of the Cosumnes River.

Overburden. Barren material, usually volcanic overlaying gold-bearing gravel deposits.

Packing the sluices. The act of running clear water through the sluices preparatory to charging or dispensing the quicksilver.

Pay dirt. A sampling of dirt containing enough gold to justify mining.

Penstock. A conduit leading water from an elevated location to the pit of a hydraulic mine. A supply line.

Piper. A hydraulic nozzle operator.

Piping. Washing a bank with any form of hydraulic nozzle.

Pounding. Guiding a stream of water from a monitor directly at the bank. This practice was discouraged in favor of a side cutting action.

Pressure box (bulkhead). A large wooden rectangular cistern placed high above a hydraulic claim which received water from a ditch, flume or pipeline. Water was led from the pressure box through a funnel-like supply line to the pit below. It was so named because the pressure or head was measured from that point to the floor of the mine.

Puddling. Preventing seepage in a ditch by slowing the flow of water, causing a buildup of silt forming a sealing effect.

Quaternary period. In geological time, the age following the Tertiary period when the present (or modern) mountains and river systems were formed.

Rifled nozzle. An interior design engineered to facilitate the flow of water from the nozzle in a solid projection. It usually consisted of three radial plates, equally spaced, inside the upper elbow of a monitor.

Riffles. Wooden bars or blocks placed in the bottom of a cradle, long tom or sluice to catch and hold the washed gold. Riffles were also made with cobble stones or iron rails.

River mining. Mining the bed of a river by diverting the flow of water.

Rocker (cradle). A rectangular box provided with a removable hopper and equipped with riffles. The box was mounted on rockers much like a baby"s cradle. The sand was dumped into the hopper and washed through the box leaving the gold behind the riffles.

Run. The period of time of washing between clean-ups.

Sand box. A receptacle in front of and attached to a pressure box positioned below the incoming flume or ditch, designed to catch any remaining sand in the water.

Shallow diggings. Placer mines in Quaternary deposits along the river bars and ravines of the present or modern rivers and streams, worked either by pan, rocker, long tom or sluice.

Slickens. The debris washed from a hydraulic mine.

Sluice. Long rectangular open-ended troughs designed to be used singly or attached to each other in a chain and equipped with riffles. By means of a stream of water, the gold was separated from the dirt.

Southern mines. A term used to describe the mines located south of the Consumnes River.

Sponge gold. After amalgam was heated, either from an open flame or in a retort, driving the mercury off, the gold that remained had the appearance of a sponge.

Stone boat. A wooden sled-like vehicle usually drawn by a horse or mule, used to remove boulders from a riverbed or the floor of a hydraulic mine.

Tail sluice. A long series of sluices set at the end of the main sluices and undercurrents, designed to salvage any remaining gold in the tailings.

Tertiary period. The geological time, encompassing the Eocene epoch, when the ancient riverbeds, containing auriferous gravel deposits were created. During the later Miocene epoch, of the Tertiary period, there followed a great outpouring of lava and volcanic ash covering these river systems and trapping their gravel deposits.

Test Shafts. Exploratory shafts sunk to ascertain the course, dimensions and yield of a Tertiary gravel deposit.

Undercurrent. A wide, shallow sluice set below the main sluice, usually at right angles. It was designed to receive finer material, passed through a grizzly, from the main sluice.

Waste gate. A device placed at regular intervals along a ditch or flume so that water could be readily turned out in the event of an accident or to prevent freezing.

Water blast. A method of ventilation used in the construction of bedrock tunnels where a current of cool air, caused by falling water, was projected through a pipe, to the workmen at the face.

Whim. A machine worked by horse, steam or water for raising ores, such as from a coyote shaft.

Wingdam. A temporary water-tight, wooden, brush or dirt dam constructed from the bank out into the stream and then downstream in the form of an L.

Bibliography

NEWSPAPERS

Butte *Record*, 1870-1873
Calaveras *Weekly Chronicle*, 1853
Columbia *Gazette*, 1856
Downieville *Mountain Messenger*, 1874
Eureka City *Sierra Citizen*, 1854
Grass Valley *Daily Union*, various issues 1865-1893
(Grass Valley*) The Hydraulic Miner*, 1939-1941
Grass Valley *National*, 1869
Marysville *Daily Appeal*, 1883-1884
San Francisco *Mining and Scientific Press*, various issues 1860-1909
Nevada City *Daily Gazette*, various issues 1866-1878
Nevada City *Democrat*, 1861
Nevada *Daily Transcript*, various issues 1861-1887
North San Juan *Hydraulic Press*, 1859
Placer *Herald*, 1855
Sacramento *Daily Union*, 1854-1857
San Juan *Times*, 1883
Trinity *Journal*, 1860

PERIODICALS

Bidwell, John, "The Discovery of Gold in California," *The Century Illustrated Monthly Magazine*, Volume XLI, No. 4, (February 1891).
Farley, Judson, "Yuba Hydraulic Mines," *Overland Monthly*, Volume V, (September 1870).
_____. "Roses Bar," *Overland Monthly*, Vol VII, (1871).
Foley, Doris, "English Dam Catastrophe Hastened Sawyer Decision," Nevada City *Nugget*, Centennial Publication, May 18, 1951.
Hittell, John S., "Mining Excitements in California," *The Overland Monthly*, Volume II, 1869.

Johnson, J. W., "Early Engineering Centers in California," California Historical Society *Quarterly*, Volume XXIX, 1950.

Loyd, Ralph C., "Gold Mining Activity in California," *California Geology*, (August 1961, Volume 34, No. 8).

Waite, E. G., "Pioneer Mining in California," *The Century Illustrated Monthly Magazine*, Volume XIII, No. 1, (May, 1891).

GOVERNMENT PUBLICATIONS

Averill, Charles Volney, "Placer Mining For Gold in California," California Division of Mines and Geology *Bulletin* 135, (San Francisco: October 1946).

Boyle, Errol Mac, "Mines and Mineral Resources of Nevada County," State Mineralogist *Report*, (Sacramento:1919).

Bradley, Walter W., *California Journal of Mines and Geology*, Quarterly Chapter of State Mineralogist *Report* 38, (San Francisco: January 1942).

Browne, J. Ross, and J. W. Taylor, *Reports Upon the Mineral Resources of the United States West of the Rocky Mountains*, (Washington: 1867).

Clark, William B. and Lydon, Philip A. "Calaveras County, California," California Division of Mines and Geology *Report* 2, 1962.

_____. "Gold Districts of California," *Bulletin* 193, California Division of Mines and Geology, (Sacramento: 1976).

DeGroot, Henry, "Hydraulic and Drift Mining," Second *Report* of the State Mineralogist of California, (September 1882).

Gardner, E. D. and Johnson, C. H., Placer Mining in the Western United States, Part II, "Hydraulicking, Treatment of Placer Concentrates, and Marketing of Gold," *Information Circular*, United States Bureau of Mines, (Washington: October 1934).

Gilbert, Grove Karl, "Hydraulic Mining Debris in the Sierra Nevada," United States Geological Survey *Professional Paper 105*, (Washington: 1917).

Haley, Charles Scott, "Gold Placers of California," California State Mining Bureau, *Bulletin* 92, (June 1923).

Hammond, John Hays, "Auriferous Gravels of California," Ninth Annual *Report* of the State Mineralogist, (Sacramento: 1890).

Hanks, H. G., "Placer, Hydraulic and Drift Mining," California Mineralogist Second *Report*, (Sacramento: 1882).

Hill, James M., "The Mining Districts of the Western United States," United States Geological Survey *Bulletin* 507, 1912.

_____. "Historical Summary of Gold Silver, Copper, Lead and Zinc Produced in California 1848 - 1926, United States Bureau of Mines *Economic Paper* 3, (Washington, 1929).

Ireland, William, Jr., "Spring Valley Hydraulic Mine," Sixth Annual *Report*, of the State Mineralogist, (Sacramento, 1886).

Jackson, W. Turrentine, "Report on the Malakoff Mine, the North Bloomfield Mining District, and the Town of North Bloomfield," Division of Beaches and Parks, Department of Parks and Recreation, 1967.

Jarman, Arthur, "Report of the Hydraulic Mining Commission Upon the Feasibility of the Resumption of Hydraulic Mining in California," A *Report* to the Legislature of 1927.

Jenkins, Olaf P., "Geology of Placer Deposits," *Special Publication 34*, California Division of Mines and Geology, (San Francisco: 1964).

Julihn, Carl E. "Mineral Industries Survey of the United States: California. Tuolumne and Mariposa Counties. Mother Lode District (South)," United States Bureau of Mines *Bulletin*, 424 (Washington, 1940).

Lindgren, Waldemar, "The Tertiary Gravels of the Sierra Nevada of California," United States Geological Survey *Professional Paper*, 73, (Washington, 1911).

Logan, C. A. and Franke, Herbert, "Mines and Mineral Resources of Calaveras County." California Journal of Mines and Geology, State Mineralogist *Report*, XXXII, (San Francisco, 1936).

_____. "History of Mining and Milling Methods in California," Geological Guidebook, California State Division of Mines, *Bulletin* 141, September 1948.

Miner, J. A. "Spring Valley Hydraulic Gold Mine," Tenth Annual *Report* of the State Mineralogist, (Sacramento, 1891).

O'Brien, J. C., "Mines and Mineral Resources of Trinity County, California," County *Report*, 4, (San Francisco, 1965).

Raymond, Rossiter W., *Mineral Resources of the States and Territories West of The Rocky Mountains*, (Washington, 1869).

_____. *Mineral Resources of the States and Territories West of the Rocky Mountains*, (Washington, 1973).

_____. *Mineral Resources of the States and Territories West of the Rocky Mountains*, (Washington, 1877).

_____. *Statistics of Mines and Mining in the Western States, and Territories West of the Rocky Mountains*, (Washington, 1873).

Report of the California Debris Commission, (November 15, 1893).

Robinson, F. W., "Notes on Hydraulic Mining," Second Annual *Report*, of the State Mineralogist, (Sacramento, 1882).

Shamberger, Hugh A., "The Story of the Water Supply for the Comstock Including the Towns of Virginia City, Gold Hill and Silver City, Nevada," United States Geological Survey *Paper*, 779, (Washington, 1972).

Smith, Hamilton, Jr., "The Bowman Reservoir and Dam," Second *Report* of the State Mineralogist, (September, 1882).

Staley, W. W., "Elementary Methods of Placer Mining," Idaho Bureau of Mines and Geology, *Pamphlet* No. 35, (Moscow, 1933).

Yeend, Warren E., "Gold Bearing Gravel of the Ancestral Yuba River, Sierra Nevada, California," United States Geological Survey *Professional Paper,* 772, Washington, 1974).

MANUSCRIPTS

Allen, Chester B., "Report of Properties of the La Grange Mining Company," Trinity County Historical Society Collection, Weaverville.

Egenhoff, Rowland I., "The Hydraulic Mining Situation in California," Read before the Sacramento Section of the American Society of Civil Engineers, January 28, 1931.

"La Grange Hydraulic Mining Co.," Trinity County Historical Collection, Weaverville.

"Rich Claim Iowa Hill, 1853," (Barber & Baker), Graphic Letter Sheet, California History Room, Sacramento.

The Milton Mining and Water Company. Manuscript Collection, California History Room, Sacramento.

Sheppard, Susan, "La Grange Mine Trinity County," Trinity County Historical Society Collection, Weaverville, CA.

PAMPHLETS

Black, George, *Report on the Middle Yuba Canal and Eureka Lake Canal, Nevada County, California,* (San Francisco: Privately Printed, 1864).

Evans, Taliesin, *Hydraulic Gold Mining in California,* (Golden, Colorado, Outbooks, 1981). Reprinted from an article published in *Century Magazine,* 1883.

Goodyear, Hal, "Giants of the Gold Rush,"*Trail Guide for the Henness-Zumwalt Pass,* E. Clampus Vitus, 1982).

_____. "Peter Minert Paulsen, " Trinity County Historical Society *Yearbook,* 1989.

Greenland, Powell, "Knight's Foundry," *The Branding Iron,* (Los Angeles: Westerners, 1976).

History and Heritage Committee, Sacramento Section, American Society of Civil Engineers, *Historic Civil Engineering Landmarks of Sacramento and Northeastern California,* (Sacramento, 1976).

Joshua Hendy Machine Works, *Hendy Hydraulic Giants,* (San Francisco, 1922).

Kallenberger, W. W., *Memories of a Gold Digger,* (Privately printed, 1970).

Litchfield, Letitia M., "Smartsville, Timbucktoo and Vicinity," (Marysville: The Yuba County Historical Commission, 1976).

Nevada County Historical Society *Bulletin* Volume 49, No. I (January 1995).

Sturgeon, Jack D., "Cherokee Flat," Butte County Historical Society Quarterly *Diggins,* Volume X:4 (Winter, 1966).

Stretch, R. H., *Iowa Hill Canal and Gravel Mines, Placer County, California,* (New York: Hopsford & Sons, 1879).

Van Der Pas, Peter W., "The English Dam Disaster, June 18,1883," Nevada County Historical Society *Bulletin*, Volume 37, No.3, (July 1983).

Wyckoff, Robert M., *Hydraulicking, a Brief History of Hydraulic Mining in Nevada County, California*,(Nevada City: Osborn, Woods, 1962).

_____. *Hydraulicking, North Bloomfield and Malakoff Diggins State Historic Park*, (Nevada City, 1993).

CONTEMPORARY BOOKS

Audubon, John W., *Audubon's Western Journal: 1849-1850, Being the MS. Record of a trip from New York to Texas, and an Overland Journey Through Mexico and Arizona to the Gold Fields of California*, Frank Heywood Hodder, ed. (Glorieta, New Mexico, The Rio Grande Press, 1969).

Bean, Edwin F., *Bean's History and Directory of Nevada County, California, Containing a Complete History*, (Nevada City, Calif.,1867).

Bowie, Aug. J., Jr., *A Practical Treatise on Hydraulic Mining in California*, (New York: D. Van Nostrand Co., 1885).

_____. "Mining Debris in California Rivers." Technical Society of the Pacific Coast, Transactions, Vol. IV (Feb. 1887).

Borthwick, J. D., *Three Years in California*, (Oakland: Biobooks, 1948).

Bowles, Samuel, *Our New West, Records of Travel between the Mississippi River and the Pacific Ocean*, (Hartford: Hartford Publishing Co., 1869).

Buffum, E. Gould, *Six Months in the Gold Mines From a Journal of Three Year's Residence in Upper and Lower California 1847-8-9*, John W. Caughey, ed.,(Los Angeles: Ward Richie Press, 1959).

California Miners Association, *California Mines and Minerals*, (San Francisco: 1899).

Clark, Walter Van Tilburg, ed. *The Journals of Alfred Doten, 1849-1903*, 3 vols., (Reno: University of Nevada Press, 1973).

Cox, Isaac, *The Annals of Trinity County*, (Eugene: University of Oregon, 1940).

Cronise, Titus Fey, *The Natural Wealth of California*, (San Francisco: H. H. Bancroft, 1868).

Decker, Peter, *The Diaries of Peter Decker Overland to California in 1849 and Life in the Mines 1850 - 1851*, Helen S. Giffen, ed. (Georgetown, California, The Talisman Press, 1966).

Ellis, W. T., *Memories, My Seventy-Two Years in the Romantic County of Yuba, California*, (Eugene: University of Oregon, 1939).

Ferguson, Charles D., *California Gold Fields*, (Oakland: Biobooks, 1948).

Gardner, Howard C., *In Pursuit of the Golden Dream, Reminiscences of San Francisco and the Northern and Southern Mines, 1849-1857*, Dale L. Morgan, ed. (Stoughton, Mass., Western Hemisphere, Inc. 1970).

Gold Mines and Mining in California, excerpts from the Mining and Scientific Press, (San Francisco: George Spaulding & Company, 1885).

Hittell, John S., *The Resources of California: Comprising, Agriculture, Min-ing, Geography, Climate, Commerce, etc. And the Past and Future Development of the State*, (San Francisco: A. Roman and Co., 1866).

_____. *Mining in the Pacific States of North America* (San Francisco: H. H. Bancroft & Company, 1861).

Hutchings, James Mason, *The Miner's Own Book*, reprinted from the original edition of 1858 with an introduction by Rodman Paul, (San Francisco: Book Club of California, 1949).

Kelly, William A., *A Stroll Through the Diggings of California*, (Oakland: Biobooks, 1950).

Nasatir, A. P., *A French Journalist in the California Gold Rush, The Letters of Etienne Derbec,* (Georgetown, California, Talisman Press, 1964).

Richards, Robert H., *Mining Engineering Notes*, (Boston, Thomas Todd, 1904).

Royce, Josiah, *California From the Conquest in 1846 to the Second Vigilance Committee in San Francisco, A Study of American Character*, (Boston: Houghton Mifflin and Company, 1886).

Sawyer, Lorenzo, *Way Sketches*, (New York: Edward Eberstadt, 1926).

Shinn, Charles Howard, *Mining Camps, A Study in American Frontier Gov-ernment*, (New York: Alfred A. Knopf, 1948).

Silliman, Benjamin, *Report on the Deep Placers of the South & Middle Yuba, Nevada County, California*, (New York: privately printed, 1865).

Wilson, Eugene, B., *Hydraulic and Placer Mining*, (NewYork: John Wiley & Sons, 1898).

Yale, Charles G. "The Mineral Industry of California," (San Francisco: California Miner's Assn., 1899).

SECONDARY BOOKS

Bancroft, Hubert Howe, *History of California*, 7 volumes, facsimile edition, (Santa Barbara: Wallace Hebberd, 1963).

Buckbee, Edna Bryan, *The Saga of Old Tuolumne*, (New York: the Press of the Pioneers, Inc. 1935).

Caughey, John Walton, *Gold is the Cornerstone*, (Berkeley:University of California Press, 1948).

Cowan, Robert G., *Ranchos of California*, (Los Angeles, 1977).

Dana, Julian, *The Sacramento, River of Gold*, (New York: Farrar and Rinehart, 1939).

Dane, G. Ezra, *Ghost Town* (New York, Knopf, 1941).

Davis, H. P., *Gold Rush Days in Nevada City*, (Nevada City: Berliner & McGinnis, 1948).

Davis, John F. *Historical Sketch of the Mining Law in California*, (Los Ange-les: 1902).

BIBLIOGRAPHY

Fariss and Smith, *History of Plumas, Lassen & Sierra Counties 1882*, repro-duction copy (Berkeley: Howell-North Books, 1971).

Gould, Helen Weaver, *La Porte Scrapbook*, (La Porte, privately printed, 1972).

Gudde, Erwin G., *California Gold Camps*, (Berkeley: University of California Press, 1975).

Hundley, Norris, Jr., *The Great Thirst, Californians and Water, 1770s– 1990s*, (Berkeley: University of California Press, 1992).

Jackson, Jake, *Tales From the "Mountaineer,"* (Weaverville, California, Ro-tary Club, 1964).

Janicot, Michel, *The French Connection or The French in Nevada County*, (Grass Valley, California, Blue Dolphin Press, 1991).

Jones, Alice Goen, ed., *Trinity County Historical Sites*, (Weaverville, California, Trinity County Historical Society, 1981).

Kelley, Robert L., *Gold vs Grain, the Hydraulic Mining Controversy in Cali-fornia's Sacramento Valley*, (Glendale: Arthur H. Clark Co., 1959).

Lavender, David, *Nothing Seemed Impossible, William C. Ralston and Early San Fran-cisco*, (Palo Alto: American West Publishing Co.,1975).

Lardner, W. B. and Brock, M. J., *History of Placer and Nevada Counties, California*, (Los Angeles: Historic Record Co., 1924).

Mann, Ralph, *After the Gold Rush, Society in Grass Valley and Nevada City, Califor-nia, 1849–1870*, (Stanford Univ. Press, 1982).

May, Philip Ross, *Origins of Hydraulic Mining in California*, (Oakland: Hol-mes Book Co., 1970).

Paul, Rodman, *California Gold, The Beginning of Mining in the Far West*, (Lincoln: University of Nebraska Press, 1947).

Scott, Edward B., *The Saga of Lake Tahoe*, (Crystal Bay, Nevada, Sierra Ta-hoe Publishing Co., 1964).

Schneider, Jimmie, *Quichsilver, The Complete History of Santa Clara County's New Almaden Mine*, (San Jose: privately printed, 1992).

Sinnott, James J., *History of Sierra County: Downieville, Gold Town on the Yuba*, Vol-ume I, (Fresno: Mid-Cal Publishing, 1977).

_____. *History of Sierra County: "Over North" in Sierra County*, Volume V, (Fresno: Mid-Cal Publishers. 1977).

Thompson & West, *History of Nevada County, California 1880*, reproduction copy, (Berkeley: Howell-North Books, 1970).

_____. *History of Amador County, California*, (Oakland, 1881).

Wagner, Jack. R., *Gold Mines of California*, (Berkeley: Howell-North Books, 1970).

Wells, Harry L. & Chambers, W. L., *History of Butte County, California, 1881*, reproduction copy, (Berkeley: Howell-North Books, 1973).

Index

Abbey, Richard: 189, 192
Accidents: 129-130, 134, 155
Allardt, G. F: 232
Allenwood, J. M: 44
Alpha (Washington Township): 54
Alpha mine: 92
Alta Foundry: 111
Altaville: 39
Amador Canal Company: 104-105
Amador County: 39, 49, 104-105, 107
amalgamation: 144
American and Knickerbocker Co: 68
American Company: 66
American Gold Field Company Ltd: 180
American Hill: 32-34, 43-44
American Hydraulic Company: 72
American Institute of Mining Engineers: 113
American Mine: 109, 141, 242, 249, 253
American River: 38, 40, 52, 74, 160, 243, 244; Middle Fork, 40; North Fork, 23, 26, 40
Anssabel, E. (chef): 172
Anti-Debris Association of Sacramento Valley: 222, 233, 243, 249, 257-259
Aqueducts: 93
Aricola, Georgius: 29
Arnott Ditch: 100
Atlantic Pacific States Telegraph Company: 206
Atwood (Attwood), Melville (mining engineer): 18
Auburn & Bear River Canal Co: 96
Audubon, John W: 24, 27
Auguste, Madam: 199

Badger Hill: 73, 92, 205
Bagdad hydraulic mine: 103
Bailey, L. L: 166
Bald Mountain Drift Mine: 31
Baldwin hydraulic mine: 105
Bank of California: 214
Banner Mine: 208
Barren, William: 200

Bayerque, R: 200
Bear River: 38, 78, 97, 115, 227-228, 230-234, 243
Bed Rock Tunnel Company: 69
Bed-Rock Mine: 205
Belcher, Judge I. S: 248-249
Bell, A. J. (mining engineer): 193
Bell, Thomas: 200, 253
Bell, V. G: 84, 253-254
Bellows, William H: 85
Bergbaufreiheit: 77
Bidwell Bar (Butte County): 15, 48, 182
Bidwell, John: 182
Big Butte Creek: 104
Big Canyon Creek: 109. 201-202, 205
Big Gap Flume: 86
Big Meadows: 100
Big Oak Flat: 86
Bingham, Mayor: 248
Birchfield: 92
Birchville: 69, 205
Birdseye Creek Gold Mining Company: 71
Blasting: 65-66, 154-155, 215
Bloody Run: 92
Bloomfield: 151
Blue Gravel Claim (Smartsville): 68, 101
Blue Lead: 96
"Blue Lead" gravel channel: 61-64, 66
Blue Point Mine: 101, 154
Blue Tent Consolidated Mining & Water Company: 85, 89, 120, 122, 222, 242, 244-245
Blue Tent: 72, 90
Board of Levee Commissioners: 228
Bolt's Hill claim (Trinity County): 163
Booster nozzle: 133; see also Nozzles
Borthwick, J. D: 26
Boss mine: 261
Bovery, Pierre (mining engineer): 176, 179-180
Bowie, August J. (mining engineer): 91, 118, 126, 259
Bowles, Samuel: 53, 76

Bowman Dam and Lake: 203-204, 207, 211, 254; photo of: 212
Bowman's Ranch: 201-202, 205
Boyce Brothers: 99
Brandy City (Sierra County): 66, 100
Brandy City Hydraulic Mining Company: 259
Bremond, Marius: 199
Bridge Camp: 171
Bridgeport Township: 198
British Columbia: 55
Brown's Hill Claim: 63
Browne, J. Ross: 34, 62, 65, 76
Buckeye Hill (Nevada County): 32, 63
Buffum, E. Gould (miner): 19, 23
Bulkhead. See pressure box
Bullard's Bar: 102
Butler, Samuel (correspondent): 224
Butte County: 38, 48, 52, 103-104, 182
Butte Creek Ditch: 189
Butte Creek: 184, 188, 191-192
Butte Ditch Company: 104-105
Butte Table Mountain Consolidated Mining Company: 183
Butterfield & Company: 200
Butterworth, Samuel: 199
Byrne, J: 249

Cadwalader, George (attorney): 231, 248-249
Calaveras County: 39, 105-106
Calaveras Hydraulic Mining & Water Company: 106
Calaveras River: 105
California Debris Commission: 224, 263
California Electrical Company: 151
California State Mining Bureau: 134
California Water Company: 94-95
Calkins, L. S: 239, 249
Caminetti Act (1893): 263, 266
Campagnie Francais de Placers Hydraulique de Junction City: 168
Campo Seco Company: 106
Camptonville: 102
Canyon Creek Lake. See Eureka Lake Reservoir
Canyon Creek: 100, 112, 168
Canyon Creek-Little Bear River Ditch: 96
Cape Horn: 101
Capital investment, Trinity County: 165
Carquinez Strait: 244
Carson Wilkinson & Co: 81-82
Cascade Lakes: 91
Case, Major: 49
Caucasian League: 190
Caughey, John Walton (historian): 17
Cave City: 106

Cedar Creek Company: 69
Cedar Grove: 100
cement mills: 62-64, 66
Central Hill Channel: 39
Central Pacific Railroad: 89, 98
Chabot, Antoine (miner): 32, 34
Chamber's Bar (Tuolumne County): 80
Champion nozzle: 121; see also Nozzles
Charity Blue Gravel Claim: 183
Cherokee (Nevada County): 69
Cherokee Creek: 100
Cherokee Flat (Butte County): 143
Cherokee Flat Blue Gravel Company: 183-184, 188; inverted siphon at, 188; merger with Spring Valley Mine, 189
Cherokee Mine: 103
Cherokee: 103, 151, 182-183, 184, 187, 189-190, 193
Chili Gulch: 49, 106
Chimney Hill: 69
China Gulch: 105
Chinese Camp: 39
Chinese labor: 81, 89-90, 108-109, 114-115, 138-139, 190, 204, 207, 210, 238, 244-245, 254; profit from Sawyer decision, 258
Chips' Flat (Sierra County): 84
City of Marysville vs. The North Bloomfield Gravel Mining Company et al: 222
Clear Creek (Shasta County): 160
Coal Creek: 169
Coldstream Canyon: 98
Collins, Daniel: 237
Coloma: 15
Columbia (Tuolumne County): 49, 75
Columbia & Stanislaus Water Company: 76
Columbia Hill: 94, 151, 198, 249, 261
Comstock Lode: 55, 187
Concow Valley and reservoir: 184, 191-192, 194
Conly & Gowell: 101
Consolidated Mining Company: 128
Coon Hollow: 95
Cornish laborers: 208
Cornucopia mine: 261
Cornwall: 77
Corps of Engineers: 235
Cosumnes River: 95
Cox, Isaac (historian): 162
Cox Pan: 64
Coyote (Calaveras County): 27
Coyote Bar: 26
Coyote Diggings (Nevada County): 25-27; sketch of, 25
Coyote Diggings (Plumas County): 26
Coyote Diggings (Tuolumne County): 27

Coyote Lead: 25
Coyote Water Company: 82
Coyoteville (El Dorado County): 27
"Coyoting" (mining technique): 25
Cradle: see Mining techniques
Craig, R. R. and J: 44, 117-119, 122
Crevicing: see mining techniques
Crum, A. J: 192
Currier, J. W: 202

D. O. Mills & Company: 75, 104
Daguerer Point: 234
Damascus: 97
Dams: 91; See also Reservoirs
Davis, Col. W. S: 152
Davis, George: 250
Davis, H. P: 78
Davis, Isaac E: 189
De re Metallica (book): 29
Debris: see Mining debris
Debris dams: 234, 237, 243, 264, 277; breaking of Rudyard Reservoir, 250-255; Calaveras County, 265; Nevada County, 265; photo of, 265
Decker, Peter (miner): 26-27, 79
Deer Creek Mining Company: 27
Deer Creek Water Company: 82
Deer Creek: 24, 27, 47, 82, 90, 92, 101, 143, 170, 243
Deflectors: 124-127
DeGroot, Henry: 33, 46
Del Norte County: 107, 163
Delaware mine: 261
Delhi mine: 261
Depot Hill Hydraulic mine: 102
Derbec: 151
Derbec drift mine: 114, 249
Derricks: 148-149, 165
Dewey, S. L: 189
Dictator nozzle: 120-121
Distributing boxes: 50; photo of, 51
Ditch tenders: 178
Ditches: 50, 52, 67, 72, 74, 80, 82, 89, 90, 92-93, 95-96, 99-101, 104-105, 111; and snow, 89; and waste gates, 82-83; and water leakage, 90; early failures of, 72-73; engineering of, 83-85; failure of many companies, 76; in Nevada County, 27; in Trinity County, 165; in Butte County, 184, 189; in Nevada County, 203, 205; list of principal, 277. See also Flumes
Dobbins, A. M: 205
Dogtown: 107
Doten, Alfred: 22, 28
Double Springs: 22

Douglas City: 161
Downie, Major: 115
Downieville: 100
Doyle, J. B: 168
Drainage Act (1880): 222, 234, 235, 238; declared unconstitutional, 239
Drainage tunnels: 67, 68, 69
Drawing (hydraulic mining term): 133
Drift mines: 68
Drift mining: 113, 156; after Sawyer decision: 258-259, See tunnel or drift mining
Drought years: 58-59
Dry Creek: 101, 105, 192
Dry diggings, defined: 15
Dunning, R. H: 141-142
Duryea mine: 262
Dutch Flat (Placer County): 54, 61, 69, 76, 96, 118, 134, 138, 231
Dutch Flat Gravel Mining Company: 96
Dutch Flat hydraulic mine: 92
Dutch Flat region: 40, 96
Dutch Flat Water Company: 96
Duty of the miner's inch, defined: 116-117
Duvergey, H: 168
Dyer Claim: 167

Eagle Bird mine: 261
Eagle Iron Works: 111
Eagle mine: 108
Edison, Thomas A: 193
Edison Company: 151
El Dorado channel: 105
El Dorado County: 39, 52, 95
El Dorado Water & Deep Gravel Mining Company: 95
Elephant Hydraulic Mine: 105, 107
Elevator: 224
Elizabeth tract: 248
Emory's Crossing: 251
Empire Foundry: 122, 236
Empire Mine: 209
Englebright, Rep. Harry L: 266
English Company: 253
English Dam. See Rudyard Dam
Eureka: 151
Eureka City (Sierra County): 48
Eureka claim: 183
Eureka Flat: 101
Eureka Hill: 62
Eureka Lake: 94
Eureka Lake & Yuba Canal Companies: 94, 109, 151, 204
Eureka Lake Company: 72, 93, 200, 207, 253
Eureka Lake flume: 251

Eureka Lake Reservoir: 93
Eureka South (Graniteville): 72
Evans, George H: 152
Evans, Taliesin: 243
Evans Elevator: 152
Excelsior Ditch: 101
Excelsior Water & Gravel Mining Company: 150, 241
Exclusion Bill (1881): 204

Fair Play hydraulic mine: 99
Farley, A. Judson: 228
Farmers: 231; lose case against miners on appeal, 233
Farms, effected by mines: 192
Farnsworth's Saloon (Columbia): 75
Faucherie, Benoit (mining engineer): 93, 149
Feather River: 38, 52, 182, 189, 191-192, 228, 230, 244, 248
Feather River & Ophir Water Company: 103
Feather River Ditch: 101
Ferguson, Charles (miner): 24, 27
Fiddle Creek: 100
Fisher, Frank: 122-123, 126
Flouring (hydraulic mining term): 136
Flumes: 23, 85-86, 88, 90, 93; illustration of, 87-88
Forbestown: 103-104
Fordyce Reservoir: 91
Forest City (Sierra County): 31
Forest Hill: 97-98
Forest Hill Divide: 97
Fort Mountain channel system: 39, 105
Fraser River gold rush: 55
Freeman's Crossing: 253
Freeman's Toll Bridge: 251
Fremont, John C: 80
French Corral: 40, 69, 73, 84, 92, 109, 138, 144, 151, 205, 249, 253, 259
French Corral Mine: 205
French Lake. See Eureka Lake Reservoir
Freres, Lazard: 232

Gardiner, Howard C: 80
Gardner's Point Claim: 144
Geeslin Ditch: 101
Gelder, Bailey & Company: 166
Georgetown: 37, 95
Georgia and Spanish Dry Diggings: 95
Giant. See Monitor nozzle
Gibsonville: 99-100
Gilbert, G. K. (geologist): 256-257
Glass, Louis: 145, 189, 192
Globe Monitor nozzle: 117-122; described, 119; illustration of, 120

Gold Flat: 27
Gold Hill (Placer County): 40, 61, 96
Gold Hill Hydraulic Mining Company: 96
Gold pans, described: 16; batea, 17-18
Gold production estimates: 262, 271, 273
Gold Run Case: 246
Gold Run Ditch: 67
Gold Run Ditch & Mining Company: 69, 92, 140, 240
Gold Run Gravels Company: 156-157
Golden Gate mine: 65, 261
Golden Rock Water Company: 86
Golden Wonder mine: 261
Goodyear's Bar (Sierra County): 48
Goosing (hyrdraulic mining term): 133
Gorman, D. L: 149
Gouge Eye Company: 66
Graniteville: 88, 207, 250-251
Grass Valley: 37, 58, 208, 249
Gray, Marshall: 78
Greenhorn Creek: 63, 245
Gregory, William: 189, 195-196
Grizzly Ditch Co: 52
Grizzly elevator: 154
Grizzly Hill Ditch: 100
Ground sluicing: 29; history of, 29; leads to hydraulic mining, 31; variations of, 30
Gudde, Edwin (historian): 198

Hagedorn, Larkin & Company: 63
Haley, Charles Scott (geologist): 37, 39, 134, 180
Hall, William Hammond (state mining engineer): 232, 234
Ham, J. C: 104
Ham Ditch: 104
Hamilton: 182
Hamilton, Bob: 254
Hammond, John Hays (mining engineer): 145, 216
Hanks, H. G. (state mineralogist): 144
Hard rock mining: 38, 55
Harry L. Englebright Dam: 268
Hart, A. L. (state attorney general): 240
Hathaway mine: 130
Hawaiian labor: 191
Haymond Bill: 231
Hendel, Charles: 143
Hendricks Mine: 103
Hendy Giant nozzle: 126, 179; photo of: 225
Heurtevant, F: 168
Hewitt Mine: 103
Hirschman & Company: 258
Hittell, John S: 45, 55, 57, 94
Hock farm: 248
Hoit, M. F: 92

Holmes claim (Trinity County): 163
Holt, G. W: 86
Honcut hydraulic mine: 103
Hoosier Ditch: 100
Horsetown (Shasta County): 49
Hoses: 43-44, 47, 50
Hoskins Deflector: 125, 127
Hoskins, Richard: 121-123, 125, 236
Howland Flats: 99-100, 258
Humboldt County: 107, 161, 163
Humboldt pit: 106
Humbug Brewery: 199
Humbug Canyon: 200-201, 249
Humbug Creek: 198
Humphrey, Isaac (miner): 18
Hundley, Norris: 81
Hungarian riffle: 135, 140, 153
Hunt's Hill (Nevada County): 54, 61, 63, 66
Hutchings, James Mason: 29, 34, 129; writes account of hydraulic mining, 33, 35
Hutchinson Reservoir: 194
Hydraulic Chief nozzle: 120-122
Hydraulic elevator: 151-154
Hydraulic Giant nozzle: 126
Hydraulic Miner's Association. See Miner's Association
Hydraulic Miners Association (1930s): 268
Hydraulic mining: evolution of: 45-46; and drainage tunnels, 67-69; and pipe, 110; debris controversy regarding, 115-116; down turn in activity, 53; early development of, 31-35; halcyon years for, 115; list of leaders, 274-276; need for venture capital, 71; revival of, in 1860s, 60-61, 76; sluices, 138, 155; techniques of, 127-129, 133-138, 140-141, 143-144, 146, 148, 152, 154, 156, 157; water supply and, 57; yield from, 157. See also ditches, and dams, 91

Igo: 107
Illumination of mines: 149-150
Inclined sluice: 154
Indian Bar: 106
Indian Creek: 102
Indian Hill (Sierra County): 40, 48
Inverted siphon: 112, 186-188
Ione (Amador County): 50, 106
Iowa Hill (Placer County): 40, 47, 57, 97, 116
Iowa Hill Canal Company: 97-98
Irishman's Bar: 79
Iron pipe: 44

Jackson: 105
Jackson, Turrentine (historian): 199
Jackson Creek: 104

Jackson Ranch: 250
Jamison Company: 47
Jarman, Arthur: 266
Jenkins, Olaf (geologist): 31, 36
Jensen, C. (mining foreman): 176
John C. Fall Ditch: 101
Johnson Ditch (Amador County): 49-50
Joshua Hendy Iron Works: 126, 133, 180
Joubert family: 102
Judson, Egbert: 184, 189, 192, 214, 253
Junction City: 162, 166, 168, 174, 181

Kallenberger, W. W: 129, 210
Kate Hayes Mine: 205
Keller, G. F. (cartoonist): 242
Kelly, William (miner): 17
Kennebec Bar: 101
Kentucky Flat: 94
Keyes, James H: 230-231
Keyes vs. Little York Gold & Water Company et al: 230, 232-233
Keyser, Judge Phil W: 222, 232, 239
Kilham Ditch: 104
King, Thomas Starr: 115, 158
Klamath Mountain area: 107
Knickerbocker hydraulic workings: 258
Knight, Samuel N: 149
Knox, W. F: 239

La Grange Deflector: 180
La Grange Ditch & Hydraulic Mining Company: 106
La Grange Mine (Stanislaus County): 106
La Grange Mine (Trinity County): 106, 127, 159, 186; engineering at, c. 1900, 174; history of, 160-181; investments in, 171; LaGrange, Baron & Baroness Ernest De , 172; photograph of tailings sluice at, 167; years of greatest productivity, 175
La Grange Placers, Inc: 180
La Porte: 58, 101
La Porte Channel: 99
La Porte-Scales-Brandy City Channel: 102
Labor relations, at North Bloomfield: 210
LaGrange, Baron & Baroness Ernest De : 166, 168, 172, 175
Lake Almanor: 100
Lake City: 198
Lake Faucherie: 93-94, 151
Lake Stirling Reservoir: 91
Lake Tahoe: 97
Lake Tahoe & San Francisco Water Company: 97
Lander's Bar: 101
Last Chance: 97
Latin American miners

Lindgren, Waldemar: 39-40, 194, 256, 264
Little Butte Creek: 103
Little Canyon Creek: 201-202
Little Giant mine: 261
Little Giant nozzle: 121-122, 125
Little York (Nevada County): 40, 54, 61-64, 71
Little York Hydraulic Mining Company: 155
Lode mines: 113
Logan, C. A: 29
London, as investment source: 71
Long Bar Ditch: 101
Long tom: described, 19
Long, W. D: 238
Long's Bar: 24, 103
Los Angeles County: 107
Loveridge, Orange M: 163-164
Loveridge Placer: 167
Lowell Hill: 259
Lyons, Mr. (lawyer): 80

MacLean Brothers: 168
Macy, C. F: 61
Macy and Martin: 117, 123
Magalia: 104
Malakoff: 249
Malakoff Diggings (Nevada County): 69, 129, 146, 152, 157, 159, 226; photos of, 219. See also North Bloomfield Gravel Mining Company
Mallory & Company: 64
Mammoth Bar: 152
Manzanita Diggings and Mine: 47, 78, 144, 156-157, 205, 242, 257
Marcelius & Maltman: 118
Mariposa: 37
Mariposa County: 39
Mariposa Creek: 15
Mariposa grant: 80
Mark's & Company: 66
Marsh, Charles: 27
Marshall, James: 182
Martell's Claim: 105
Martindale Ditch: 49, 101
Marysville: 122, 228, 230, 233, 235, 240, 244, 248, 252; boycott by miners against, 236, 240-241; damage from dam break, 253
Marysville Foundry: 119
Marysville Levee Commission: 229
Marysville vs. The North Bloomfield Gravel Mining Company et al: 239
Massassauga Company: 62-63
Matteson, Edward Eddy: 33-35, 43, 48, 148
May, Philip Ross (historian): 33
Mayhew, Judge H. A: 240
McCarthy Claim: 167

McClatchy, James (newspaperman): 249-250
McClure, Col. William: 47
McDonald, Richard (miner): 129
McEachern, Thomas (mining superintendent): 130
McGillivray, Joseph: 112
McGowell Ditch: 100
McKillicum's mine: 251
McLaughlin, Maj. Frank: 193
Meadow Lake Reservoir: 91
Mendell, Lt. Col. G. H: 235, 243; report of, 246
Menona Flat: 97
Messerville: 162
Michigan Bluff (Placer County): 97-98, 117, 148
Middle Fork of the American River: 94-95, 97, 152
Middle Fork of Yuba River: 51-52, 82, 101, 109, 250, 252
Middle Yuba Canal Company: 52, 60, 92-93
Mill Creek: 168
Miller, Eli: 34
Miller, N. C: 250
Mills, D. O: 214
Milton: 90
Milton ditch (Nevada County): 88, 109
Milton Mine: 242
Milton Mining & Water Company: 84, 108-109, 125, 138, 144, 151, 184, 204-205, 214, 235, 250, 252, 254, 257; absorbed by North Bloomfield Gravel Mining Co, 220
Milton Tunnel: 150
Miner's Association: 231-232, 243-244, 249, 256
Miner's Foundry. See Nevada Foundry
Mineral lands, sale of restricted: 57
Mineral Slide Mine: 104
Miners: transient character of, 57; new skills required of, 131; personal income for, 114; in Trinity County, 132
Miners' Cosumnes & Deer Creek Canal Company: 74
Minersville Hydraulic Gold Mining Company: 168
Mining codes. See Mining rights and laws
Mining debris: 227-28, 230; amount in rivers (1878), 244; effects San Francisco Bay, 244; estimates, 256-257; report on, 234. See also Debris dams
Mining engineers, new skills demanded of: 131
Mining rights and laws: 77-80, 82
Mining techniques: crevicing (knife-mining), 18; hard rock, vein, or quartz mining, 38; long tom, 19; river mining, 22-23; rocker or cradle, 18-19; sluice box, 20
Miocene Ditch (Butte County): 85

INDEX

Miocene Mining Company (Butte County): 103, 240
Mississippi Bar: 74
Mokelumne Ditch & Water Company: 106
Mokelumne Hill: 39
Mokelumne River: 105
Monitor nozzle: 61, 96, 101-102, 117-119, 166, 172, 184, 188, 203, 223, 236. See also Nozzles, Globe Monitor nozzle
Mono County: 107
Monoville: 107
Monroe, Kirk: 224
Monumental mining claim: 50
Mooney Flat: 101, 151
Moore's Flat: 78, 88, 108-109, 249
Morgan, Jenkin (miner): 142
Morristown: 72
Mosquito Creek: 27
Mother Lode: 37
Motto pit: 106
Moulton, John (mine watchman): 138
Mountain Ranch: 106
Muletown (Amador County): 50
Mull & Co: 79-80
Myers, J. R: 201

National aqueduct: 93
National Exchange Hotel (Nevada City): 71
Navigation, on California rivers: 248
Neff, J. H: 244
Neice & West Mine: 63
Nelson Creek-Rich Bar region (Plumas County): 49
Nevada City: 24, 26, 32-33, 44, 47, 78, 107, 114, 119, 148, 198, 208, 227, 230, 240, 247; meeting of miners in (1881), 237; reacts to Sawyer Decision, 255
Nevada County: 26, 39-40, 52, 54, 60, 64, 69, 72, 82, 94, 109, 115-116, 151; after Sawyer Decision, 260
Nevada County and Sacramento Canal Company: 82
Nevada Foundry: 118-120, 125
Nevada Mine: 69
New Almaden Quicksilver Mining Company: 199-200
New York, as source of venture capital: 72
North Bloomfield Gravel Mining Company: 69, 108-109, 118, 122, 125, 150-151, 160, 242, 246, 253, 259; history of, 198-226; Marysville files against, 233, 235, 240
North Bloomfield: 56, 61, 72, 90, 94, 114, 117, 124, 135-136, 143, 157; description of town, 210; fire in, 213; population swings in, 220

North Columbia: 128
North Fork Company: 100
North Fork of the American River: 97, 246
North Fork of the Feather River: 100, 107, 109
North Fork of the South Yuba: 91
North San Juan: 65, 92, 109, 151, 249
North Star Mine: 209
North Yuba River: 99
Noyes, Allen S: 195
Nozzles: 43, 133, 145, 155, 168; improvements in, 117-127; gooseneck, 44-45, 61; rifled nozzle, 61. See also Booster, Champion, Dictator, Globe, Hydraulic Chief, Hydraulic Giant, Little Giant Monitor, Watson Hydraulic
O'Brien, J: 249
Odd Fellows: 206
Ohleyer, George: 248
Old Gulch: 106
Old Star Company: 60
Omega: 139
Omega Ditch: 67
Omega Hydraulic Mine: 92, 148, 258, 264
Ophir hydraulic mine: 103
Oregon Creek: 253
Oregon Gulch: 162-166, 174; first mention of hydraulic mining at, 162
Oregon Mining District: 166
Oregon Mountain: 166, 171
Orion Mine: 116
Oroville: 103, 182
Oroville Mining & Irrigating Company: 103
Ousley's Bar: 101

Pacific Foundry: 111
Pacific Gas & Electric Company: 112
Pactolus: 101
Park Canal & Mining Company: 95
Paul, Rodman (historian): 16, 29, 31, 46
Paulsen, Peter M: 161, 163
Payne, John: 33
Pease, Mr. (Judge): 80
Pelton, Lester Allen: 149
Perkins, Henry (mining superintendent): 124-125, 222, 249, 253
Perkins Deflector: 124-125
Phelps Hill: 27
Pichoir, Henry: 84
Piety Hill Blue Gravel: 107
Pike City: 156
Pine Grove: 100, 105
Pioche, Francois Louis Alfred: 199
Pioche, Bayurque & Company: 104
Pipe: 110-112
Pipers, description of work: 132

Placer County Canal: 96
Placer County: 39-40, 52, 67, 97-98, 262; after Sawyer Decision, 261
Placerville: 95
Pleasant Flat: 90
Pliny the Elder: 29; description of sluicing: 278-279
Plum Valley: 156
Plumas County: 49, 100
Plumas Water & Mining Company: 100
Plymouth: 105
Poker Flat: 99
Polar Star Mine: 134, 138, 147
Pomeroy, Prof: 248
population; decline in mining counties: 113
Poquillon, Jules: 199-200
Port Wine: 99
Portuguese labor: 190
Potosi: 100
Poverty Hill: 99
Poverty Hill-Brandy City district: 99
Powers Mine: 103
Pressure box (bulkhead): 145-147
Pullium, R. C: 189, 192

Quaker Hill: 32, 40, 63, 72, 232
Quartz mines: 38, 55, 115
Quicksilver: 136, 143-45; use with sluice box: 21
Quitry, Chaumont (engineer): 170

Rabbit Creek (La Port) (Plumas County): 49, 99
Rabbit Creek Flume: 101
Radford, William H: 168-169
Ralston, William: 200, 213-214
Rancheria Creek: 105
Rancho Arroyo Chico: 182
Raymond, Rossiter W: 32, 40, 54, 94, 96, 107, 113, 163, 182, 192, 265
Reading, Maj. Pearson B: 160-161
Reading's Bar: 161
Red Dog (Nevada County): 40, 54, 58, 61, 117
Red Hill claim (Trinity County): 163
Red Hill Hydraulic Company: 104
Redding Creek Mine: 153
Reese River mines: 55
Relief Hill (San Juan Ridge): 54, 94
Remington Hill: 259
Reservoir Hill: 112
Reservoirs: 91. See also Debris dams
Richard, Jenkin W: 117
Ridge Telephone Company: 151
Riffle plates: 135; illustration of, 137
Risdon Iron Works: 111, 152, 186-187
Robinson, F. W: 201

Robinson, L. L: 200, 232, 249, 253
Rock Creek: 27
Rock Creek Water Company: 34
Rogers, John (miner): 26
Roosevelt, Theodore: 256
Rose Hill pit: 106
Rose's Bar: 101, 228-229
Round Valley Reservoir: 194
Royce, Josiah: 79
Ruble elevator: 153
Rudyard Dam (English Dam):109, 200, 205; breaking of, 250-255
Rush Creek: 170-71

Sacramento Bee, supports farmers: 241-242
Sacramento River: 15, 227-228, 244, 248
Sacramento Valley: 230
Sailor Flat: 90
St. Louis (Sierra County): 99
St. Louis-Table Rock region: 99
Saladin, E: 168
Salmon Mountain Range: 161
San Andreas: 39
San Francisco Board of Trade, debris committee of: 240, 243
San Francisco: 98, 111
San Gabriel Mountains: 107
San Joaquin River: 15, 228
San Juan Ridge District: 40
San Juan Ridge: 40, 51, 56, 60, 89, 93, 108, 159, 198, 261, 266
Sargent & Jacobs: 72
Saw Hill: 101
Saw Mill Bar: 101
Sawmill Canyon: 184
Sawmill Ravine: 193
Sawyer, Judge Lorenzo: 25, 122, 222, 247, 249, 256
Sawyer Decision (1884): 46, 102, 106, 126, 151, 159, 166, 175, 195, 222-223, 255, 257, 260, 262
Scales Diggings: 99, 258
Schussler, Hermann: 187
Scotchman Creek: 264
Scott Mountain Range: 161
Scotts Flat (Placer County): 67
Searls, Judge Niles: 249
Sears Ravine Flume: 142
Sears Union Water Company: 99
Sebastopol (Nevada County): 66, 68, 92, 253
Secret Treasure mine: 261
Seizer (boat): 248
Selby Hill Mine: 146
Selfridge, J. M: 165

INDEX

Serpentine belt: 39
Shands: 151
Shasta County: 49, 107, 160
Sheep Ranch: 39, 106
Shinn, Charles: 78
Shirly, Adam: 90
Sierra County: 38, 52, 81, 98-100, 109, 156
Sierra Nevada Mining & Water Company: 200
Silliman, Benjamin: 52, 71
Silversmith, J. (publisher): 141
Sinnamon Cut: 107
Sinnott, James (historian): 99
Siskiyou County: 107, 163
Skidmore, Walter A: 54, 232, 249
Slate Creek: 40, 99, 100
Slater, Mr. (lawyer): 80
Sluices: 115, 135-36, 139, 142, 144, 154; described, 20-22
Smartsville (Yuba County): 68-69, 101, 121, 236, 238, 240, 252
Smartsville ditch: 90
Smartsville Hydraulic Mining Companies: 101
Smith & Low Foundry: 51
Smith & Ringwood: 170
Smith, Hamilton, Jr: 203-205, 208-212, 214, 231-232, 249, 253-254; leaves for South America, 223; letter re Miner's Association, 280-282
Smith-La Grange deflector: 127
Snow Point: 40
Sonorans. See Mexicans
South Feather River Company: 103
South Feather Water & Union Mining Company: 104
South Fork of the American River: 94-95, 107, 109
South Fork of the Feather River: 99
South Fork of the Tuolumne River: 86
South Fork of Yuba River: 52, 79
South Fork to the Bear River: 92
South Steep Hollow: 73
South Yuba: 92
South Yuba Canal Company: 58-59, 73, 82, 89, 91-92, 96, 107-108
Southern Cross Mine: 146
Southern Mines, effect of Sawyer Decision on: 261
Spanish Gulch (Amador County): 28
Spring Valley Hydraulic Mine (Butte County): 143, 145, 159, 160, 183; employment at, 191; history of, 182-197; illustration of, 185; inverted siphon constructed at, 186-188; merger with Cherokee Company, 189; purchases farmlands, 193. See also Cherokee Mine

Spring Valley Mining & Canal Company: 189
Spring Valley Mining & Irrigation Company: 184
Spring Valley Water Company: 103, 196
Spring Valley Water Works: 187
Squaw Creek (Shasta County): 49
Staley, W. W. (mining engineer): 17
Stanislaus County: 106
Steel rails, used in sluices: 176, 178
Steep Hollow Ridge: 92
Stidger, Judge O. P: 251
Still, W. G: 81
Stockton Hill (Calaveras County): 49
Stokes, David (mining foreman): 124
Stookey, Dr. Byron: 180
Stookey, Dr. Lyman: 181
Stovall, Dennis: 132
Stretch, R. H. (mining engineer): 98
Stuart's Fork: 168, 170-171, 178-179
Sucker Flat: 97, 101, 155
Sucker Flat Channel: 101
Sugar Loaf Crest: 192
Sugar Loaf: 183
Sutter Canal & Mining Company: 104
Sutter County: 230
Sutter Creek: 50, 149
Sutter Iron Works: 111
Sutter, John: 160, 182, 248
Sweetland: 92, 109, 151, 253, 257

Table Mountain: 104, 159, 182, 184
Table Mountain Channel: 39
Table Mountain Consolidated hydraulic mine: 192
Tail sluices: 142-144
Tailings: 142-143
Tehama County: 240
Telephones: 250; and hydraulic mining, 151
Temple, Judge Jackson: 246-247
Tertiary gravel deposits: 38-40
Texas Creek: 212
Texas Hill: 74, 95
Texas Point Mine (Los Angeles County): 107
Texas Springs (Shasta County): 49
Thomas & Company: 27
Thompson Flat: 103, 189
Timbuctoo (Yuba County): 44, 60, 101, 110
Tip-Top mine: 261
Todd's Valley: 40, 97
Tomlinson Diggings. See Manzanita Diggings
Trinidad Bay: 161
Trinity Alps: 170
Trinity County: 118, 127, 161-163, 166, 172
Trinity Gold Mining Company: 165-166, 169
Trinity River: 107, 112, 160-163, 166
Tri-Union Ditch: 101

Troy Mine: 205
Truckee River: 97
Tunnel drilling: 150
Tunnel or "drift mining": 30, 31, 209, 216-217. See also Drainage tunnels
Tuolumne County: 39, 49, 75, 86
Tuolumne County Water Company: 75
Tuolumne River: 106
Tuolumne Water Company: 49
Turner's Bar: 161, 163

Undercurrent (hydraulic mining technique): 140, 142-144, 215
Underwood, Col. Joseph: 264
Union Gravel Mining Company: 204
Union Hotel (Nevada City): 78
Union Iron Works (San Francisco): 110-111, 152
Union Water Company: 100, 106
United States Supreme Court: 232

Vallecito: 39
Van Eman Brothers: 148
Vein Mining: 38. See also Hardrock mining
Venture capital: 71, 74
Verge, Eugene (mining engineer): 200
Virgin Valley Creek: 206
Virginia City, Nev: 56
Virginia Creek: 107
Volcano: 50, 105
Von Schmidt, Col. Alexis Waldemar: 97-98, 232
Vulcan Foundry: 111

Waite, E. G. (miner): 19
Waldeyer, Charles: 32, 183, 187, 194
Walkenshaw mine: 108
Walker & Wilson Ditch: 103
Wallace, Judge: 249
Walls Diggings: 74
Ward, James: 162, 167
Ward mine: 164-166
Washington: 203
Washington belt: 39
Washington Township: 67
Wasteweir Reservoir: 194
Water consumption: 107-109
Water rights: 47, 77, 80-82
Water: availability of, 46, 53, 57-58, 116; brought to Spring Valley, 186-188 cost of, 28, 52; distributing system for, 145; lack of, at Trinity County mines, 163-164, 166; means to turn off, 147; measurement of, 28; sources for Trinity County, 168; system at the La Grange, 176
Waterwheels: 149
Watson Hydraulic Champion Joint and Nozzle: 120
Watt: 151
Wattles, B. C: 163
Weaver Creek: 164
Weaver Lake: 151
Weaverville: 162-163, 181
Weaverville Ditch & Hydraulic Mining Company: 163-164
Webb, Louis: 180
Weir: 83
Weiss, Emile: 199
Welch Company: 69
Welsh Company Mine: 183, 192
West Branch of the Feather River Company: 103
West Point: 39
West Weaver: 170
Wheatland: 230, 242
Whisky Diggings: 99
White Pine mines: 55
Willow Springs Ditch: 104
Wilson, S. M. (attorney): 249
Winham, Marcos A: 51
Wisconsin Hill: 40, 97
Wolf Creek: 81-82
Wood, Joe: 32
Woodruff, Edward: 222, 247-248
Woolsey's Flat: 198
Wyandotte & Feather River Water Company: 103
Wyandotte hydraulic mine: 103

Yale, Charles G: 113
Yankee Hill: 184
Yankee Hill Ditch: 101
Yankee Jim's (Placer County): 32, 40, 43, 47, 97-98
You Bet (Nevada County): 54, 61, 69, 71, 245
You Bet Mine: 64, 144
Young America: 101
Yuba City: 233
Yuba County: 38, 52, 101-102, 110, 259
Yuba mine: 261
Yuba River: 24, 38, 40, 69, 101, 222, 227-228, 230, 234-235, 243-244. See also Middle Fork of, South Fork of
Yuba Tunnel: 68